Hakon A. Sommerfeldt

Elementary and Practical Principles of the Construction of

Ships for Ocean

and River Service

Hakon A. Sommerfeldt

Elementary and Practical Principles of the Construction of Ships for Ocean *and River Service*

ISBN/EAN: 9783337317607

Printed in Europe, USA, Canada, Australia, Japan

Cover: Foto ©berggeist007 / pixelio.de

More available books at **www.hansebooks.com**

ELEMENTARY

AND

PRACTICAL PRINCIPLES

OF THE

CONSTRUCTION OF SHIPS

FOR

OCEAN AND RIVER SERVICE.

BY

HAKON A. SOMMERFELDT,

Surveyor to the Royal Norwegian Navy ; Knight of the Order of St. Olaf.

WITH AN APPENDIX OF FURTHER INVESTIGATIONS,

AND

A LARGE ATLAS OF ENGRAVINGS DRAWN TO A
SCALE FOR SERVICE,

WHICH IS SOLD SEPARATE, OR TOGETHER WITH THE TEXT.

LONDON :
JOHN WEALE, 59, HIGH HOLBORN.
1860—61.

LONDON :

BRADBURY AND EVANS, PRINTERS, WHITEFRIARS

PRACTICAL CONSTRUCTION OF SHIPS.

CHAPTER I.

INTRODUCTION.

§ 1. *To calculate the Areas of Plane Figures.*

1. THE area of a triangle *a b c* (Pl. I., fig. 1) is equal to the basis *ab* multiplied by half the perpendicular height $cd = \frac{1}{2} ab \times cd$.

2. The area of a parallelogram *a b c d* (Pl. I., fig. 2) is equal to the basis *ab* multiplied by the perpendicular height $ef = ab \times ef$.

If $ab = bc = cd = da$, and each angle 90 degrees, the figure is a square, and the area equal to $ab \times bc = ab^2$.

3. The area of a trapezoid *a b c d* (Pl. I., fig. 3) where *ab* and *cd* are parallel, is equal to $\frac{ab + cd}{2}$ multiplied by the perpendicular height $ef = \frac{ab + cd}{2} \times ef$.

4. The area of a trapezium *a b c d* (Pl. I., fig. 4), where none of the sides are parallel, is found by dividing it into two triangles, calculating the area of each triangle separately, and taking the sum of these areas.

5. If the diameter of a circle is d and radius $=r$, d is $=2r$, and the ·area of the circle equal to $\dfrac{\pi d^2}{4}=\pi r$ $=\dfrac{d^2}{1\cdot273}=0\cdot7854d^2$, for $\pi=3\cdot1415926\ldots$ or $\tfrac{355}{113}$ or $\tfrac{22}{7}$ nearly.

The periphery or circumference of the circle is equal to $\pi d=2\pi r$.

6. If the equation of a curved line is $y^n=px$, the area Z is $=\dfrac{n}{n+1}xy$, and the exponent $n=\dfrac{z}{xy-z}$. If the exponent n is $=$ unity, the line is a straight one, and if $n=2$, the line is a common parabola.

7 The area of a plane figure A B C (Pl. I., fig. 5) included between a straight line B C, and a fair curved line B E A D C, may be found in several ways with sufficient accuracy for all practical purposes. The following formula is easily remembered, and is capable of giving the area as accurate as may be desired. Divide the straight line B C into an even number of equal parts— for instance, as in fig. 5, 6 parts; draw through the dividing points 1, 2, 3, 4, 5, lines perpendicular to B C. Let the lengths of these perpendicular lines, from the straight line B C to the curved line B E A D C, be $a\,b\,c\,d$ and e, and the perpendicular distance between them be m. Then the area is $=(1a+2b+4c+2d+4e)\dfrac{m}{3}$, and of the plane figure 1, 3, A E, the area is $=(a+4b-c)\dfrac{m}{3}$.

The more parts that the straight line в с is divided into, the more accurate will be the result, and the formula is always the same ; that is, the perpendiculars or ordinates are multiplied by $1-4-2-4-2-4-1$, beginning with $1-4$, and ending with $4-1$; as, for instance, with 9 ordinates $(a + 4b + 2c + 4d + 2e + 4f + 2g + 4h + i)$ $\frac{m}{3}$.

If the curved line has a considerable curvature it is necessary to use more ordinates than when the curvature is only slight, in order to get an accurate result ; and if the curved line in any part has an abrupt curvature, it may be preferable to divide the figure in that place into two, and calculate the areas of the two parts separately.

§ 2 To find the Cubical Contents of different Bodies.

1. The solid content of a cube whose side is $=a$, is $=a^3$.

2. The solid content of a cylinder or prism is equal to the area of the basis multiplied by the perpendicular height.

3. The solid content of a pyramid or cone is equal to the area of the basis multiplied by one-third of the perpendicular height.

4. The solid content of a ball whose diameter is $=d$, and radius $=r$, is $=\frac{1}{6}\pi d^3 = \frac{4}{3}\pi r^3$.

The superficial area of the ball is $=\pi d^2 = 4\pi r^2$.

5. $a\, b\, k\, i$ (Pl. I., fig. 6) is an irregular body, cut

through with equidistant parallel planes *ab*, *cd*, *ef*, *gh* and *ik*, whose areas are equal to A, B, C, D and E, and the perpendicular distance between these equidistant planes is $= m$. Then the solid content of this body is

$$= (\text{A} + 4\text{B} + 2\text{C} + 4\text{D} + \text{E}) \frac{m}{3}.$$

It will be observed that this formula is the same as that used for calculating the areas of plane figures (§ 1, No. 7), only the areas of the sections are here put instead of the ordinates. Consequently the same rules must here be observed in order to obtain a sufficiently correct result.

§ 3. *To find the Centre of Gravity of Plane Figures.*

1 The centre of gravity of a triangle *a b c* (Pl. I., fig. 7) is found by dividing the two sides *ab* and *bc* in half in *d* and *e*, and drawing the lines *ae* and *cd*. The point *f*, where these two lines intersect each other, is the centre of gravity.

2. The centre of gravity of a trapezoid *a b c d* (Pl. I., fig. 8), whose sides *ab* and *cd* are parallel, is found in the following manner: Divide each of the four sides in half in *efg* and *h*, draw the line *eg* between the middles of the two parallel lines *ab* and *cd*, and the lines *ed*, *bh*, *fd* and *bg*, join the intersecting points *i* and *k* with a straight line, and the point *l*, where this line intersects the straight line between the middles of the two parallel sides, is the centre of gravity.

3. In a trapezium *a b c d* (Pl. I., fig. 9), when none

of the sides are parallel, the centre of gravity is found thus: Divide each of the four sides in half in $e f g$ and h, draw the straight lines af, ag, ch and ce, and join the intersecting points i and k with a straight line ik. Draw the line bg, bh, dc and df, and join the intersecting points l and m with a straight line lm. The point n, in which lm intersects ik, is the centre of gravity.

4. In the plane figure A B C D (Pl. I., fig. 10), whose area, according to the formula in section 1., No. 7, is $=(a+4b+2c+4d+e)\frac{m}{3}$. The perpendicular distance from the centre of gravity to the line AD is=

$$\frac{0 \times a+1 \times 4b+2 \times 2c+3 \times 4d+4 \times e}{a+4b+2c+4d+e} \times m.$$

§ 4. To find the Centre of Gravity of Solid Bodies.

1. In a pyramid or cone the place of the centre of gravity is ¼ of the perpendicular height from the basis into a straight line from the apex to the centre of gravity of the basis.

2. In the irregular body $a b k i$ (Pl. I., fig. 6), whose solid content is $(A+4B+2C+4D+E)\frac{m}{3}$ (section 2, No. 5), the perpendicular distance from the centre of gravity to the plane ab is=

$$\frac{0 \times A+1 \times 4B+2 \times 2C+3 \times 4D+4E}{A+4B+2C+4D+E} \times m.$$

3. The bodies A, B, C, whose solid contents are equal,

A, B and C, are so situated that the centre of gravity of
A is in a, of B in b, and of C in c (Pl. I., fig. 11). Suppose their common centre of gravity to be in d. If
the perpendicular distances ah, bi and ck from a plane
fg are known, the distance dl or the perpendicular
distance from the common centre of gravity d, to the
same plane fg is $= \dfrac{A \times ah + B \times bi + C \times ck}{A + B + C}$, always
under condition that the weights of the bodies A, B, and
C, are proportional to their solid contents.

§ 5. On the Metacentre and Stability of Floating Bodies.

When a body A B C (Pl. I., fig. 12) whose centre of
gravity is in a is thrown into the water, it will lie
quiet only when the point a and the centre of gravity
of the displaced water b are in the same vertical line
D C, in such a manner that the distance between these
two points is the shortest possible.

If a force, for instance the wind in the sails of a
vessel, acts on the body in a manner that its position
in the water is altered, so that the line G H (fig. 12)
becomes the water-line instead of the line E F, the
centre of gravity of the body is still in the same place
a, but the centre of gravity of the displaced water is
carried to another place, say to c. If a straight line is
drawn from c perpendicular to the new water-line G H,
and produced till it intersects the original vertical line
C D in the point B, this point B is called the metacentre.

When the metacentre, as in fig. 12, is above the centre of gravity a of the body, the body will have a tendency to go back to its original position ; but if, as in fig. 13, Pl. I, the metacentre R is below the centre of gravity a of the body, the body will instantly upset or roll over on one of its sides, without any tendency to go back to its former position. Consequently it is impossible that a body, thrown into the water, can by itself, without any external support, take a position in which the metacentre is below the centre of gravity of the body.. By this we learn that it is indispensable that a ship shall be so formed and loaded that the metacentre under all circumstances comes a considerable distance above the centre of gravity of the ship.

Of two floating bodies, whose solid contents, specific gravity, and depth below the water are the same, the one that is the broadest at the water-line has its metacentre highest and is the stiffest.

When two floating bodies are homogeneous and homologous, and their breadths are B and b, their stabilities are to each other as B^4 to b^4. If two homogeneous bodies have homologous transverse sections, but not homologous longitudinal sections, and their lengths are L and l, and breadths B and b, their stabilities are to each other as $L \times B^3$ to $l \times b^3$.

By this it is seen that it is principally the breadth that produces stability, and a rough comparison between the stabilities of two ships may be made by comparing the products of their lengths and cubes of their breadths.

Of two homogeneous bodies whose lengths, breadths, and depths, and areas of transverse sections, consequently also the solid contents, are equal, the body is the stiffest in which the centre of gravity of the displaced water is nearest to the water's surface.

A B C (Pl. I., fig. 14) is half the upper or load water-line of a ship, the lines a, b, c, d, e equi-distant and perpendicular to the line A B, and the distance between them $= m$. The content of the displaced water of the ship is D. Then the metacentre for an infinitely small angle of inclination or deviation from the upright position is above the centre of gravity of the displaced water for the part D E G F,

$$\frac{\frac{2}{3}\left(a^3 + 4\,b^3 + 2\,c^3 + 4\,d^3 + e^3\right)\frac{m}{3}}{D}$$

for the triangle A D E, $\dfrac{\frac{2}{3}\,AD \times DE^3}{4D}$

for the triangle B G F, $\dfrac{\frac{2}{3}\,BF \times FG^3}{4D}$ and

for the whole body,

$$\frac{\frac{2}{3}\left(a^2 + 4\,b^3 + 2\,c^3 + 4\,d^3 + e^3\right)\frac{m}{3} + \frac{AD \times DE^3}{4} + \frac{BF \times FG^3}{4}}{D}$$

If there are more ordinates their coefficients are, as before, $1 - 4 - 2 - 4 - 2 - 4 \ldots 4 - 1$.

Supposing that in ships generally the centre of gravity is in the load water-line, or, in proportion to their size, near this water-line, which is nearly true; the distance that the metacentre is [above the load water-line multiplied by the weight of the ship may be considered as a measure of the stability. However, this will not hold good for full and deep merchant ships, in which the centre of gravity necessarily by the lading must be brought considerably below the load water-line, that they may acquire the necessary stability.

CHAPTER II.

§1. On the Displacement.

THE volume of water which a ship displaces, when lying in undisturbed water, is called the displacement of the ship. The weight of the water which the ship displaces is equal to the weight of the ship, with every-thing that is in it.

When a drawing for a ship is to be made, it is necessary to have the displacement calculated in cubic feet. The weight of one cubic foot of sea-water may be estimated on an average to 64 lbs. avoirdupois, and consequently the weight of 35 cubic feet of sea-water is equal to one ton.

When a ship is to be built for a given trade or a man-of-war is to carry a certain number of guns of a given size; the weight of the cargo that the ship is to be able to carry, or the weight of the man-of-war's ordnance, men, provisions, stores, steam-engines, coals, rigging, and sails; in short, the weight of everything that is to come into her, is calculated in tons, and to this weight is added the weight of the ship's hull.

It is impossible before a complete drawing is made to calculate the weight of the ship's hull exactly; but for merchant vessels, in which the weight of the crew,

with provisions and sea-stores, is inconsiderable in proportion to the weight of the hull and cargo, the weight
of the ship with rigging, sea stores, crew, and provisions may be found with the necessary degree of
exactness by putting the weight of the ship, with every
thing in it, except the cargo $= f_L$, where L is the weight
of the cargo. To determine the coefficient f, it must
be observed that a flat-bottomed full vessel will be
lighter in proportion to the cargo that she is able to
carry than a sharp one, and therefore f must be greater
for sharp than for flat-bottomed vessels. If the greatest
breadth of the load water-line is $=$ B, and the depth
from the load water-line to the rabbat of the keel is
$= d$, B × d is the area of a parallelogram circum-
scribing the greatest vertical transverse section of the
ship's immersed body, and the area of this transverse
section may be put $= m \times$ B $\times d$. The quantity m deter-
mines in a considerable degree the sharpness of the ship's
body, as will be seen hereafter, and therefore the coefficient f may be put $= \dfrac{x}{m}$. By experience it is found
that x for all common merchant vessels is $=$ one-half
$= 0.5$ nearly; therefore the weight of the vessel,
with everything in it except the cargo, is $= \dfrac{0.5}{m} L$, and
the whole displacement $= L + \dfrac{0.5}{m} L$.

For uncommonly sharp and strong-built ships, and
for those destined for long voyages, that are obliged to
take in a larger stock of provisions and sea-stores, x

must be put $= 0.6$, and the whole displacement $=$
$L + \dfrac{0.6}{m} L$.

Men-of-war are now always provided with steam power, greater or less, according to the speed intended to be attained, and therefore generally built with finer lines than most sailing merchant ships; consequently their hulls are heavier in proportion to what they are able to carry than the hulls of merchant vessels. If the weight of everything that comes into a man-of-war (except masts, spars, rigging, sails, and sea-stores), such as guns with carriages, ammunition with gunners' stores, crew with water and provisions, steam-engines with boilers, water in boilers and coals, and ballast, if the ship is to have any, is called s, the weight of the ship's hull may be put equal to from 1·25 s to 1·5 s, according to the intended sharpness of the ship : 1·25 s to he used when the ship is to be rather full for a man-of-war, and 1·5 s when it is to be very sharp or is to have very fine lines. The weight of the masts, spars, rigging, sails, anchors, cables, boats, boatswains' and carpenters' stores, will, in ships with the largest steam power, be nearly equal to 0·1 s, and in ships with a small auxiliary steam power about 0·3 s. The whole displacement will, therefore, be from s + 1·25 s + 0·1 s to s + 1·5 s + 0·3 s.

The displacement calculated in this manner in tons, multiplied by 35, gives the displacement in cubic feet. This displacement is considered to be on the timbers or frames without the planking, for the specific gravity of the planking with bolts and copper sheathing will be nearly the same as of sea-water. Still the displacement calculated in this manner will only come near to what it actually will be, in consequence of the uncertainty in the supposed weight of the hull and rigging, and it is, therefore, necessary, when the principal dimensions of the ship are calculated, to compare the displacement with that of ships of similar dimensions, already built of the same description of materials. This comparison will decide if any alterations are to be made in the supposed weight of the hull, rigging, &c., and consequently also in the calculated dimensions.

The displacement of merchant vessels with steam power, intended chiefly to carry goods, may be calculated in the same manner as for sailing-vessels, only it is to be observed that the weight of the steam-engine with boilers, water in boilers and coals must be added to the weight of the cargo that the ship is intended to carry, and the sum of these weights considered as the weight of the cargo for which the ship is to be constructed.

Steam-ships for passengers, built with fine lines to obtain great speed, are something heavier, and their hulls will be nearly of the same weight as of that which the ship is intended to carry, engines, coal, passengers, and everything included.

These rules will hold good for sea-going vessels built of oak. Iron ships will be from 10 to 15 per cent. lighter, and river steam-boats still lighter.

If it is desirable to know the draught of water of such vessels accurately beforehand, as it may be for some river boats, the drawing must be made up complete, and then the weight of the hull calculated according to the materials intended to be used and their dimensions, and thereafter such alterations made as may be found necessary to procure the required displacement.

For the calculation of the displacement for a man-of-war, the following will be useful.

One 68-pounder gun of 95 cwt., with 150 charges, carriage
   breach, tackles and reserve pieces, weighs . . . 12·5 tons
The same with all appurtenances but only 80 charges , . 10·0 „
One 8-inch gun of 65 cwt. with 80 charges and appurtenances 7·2 „
One 32-pounder of 56 cwt. „ „ „ „ 6·0 „
 „ 32 „ 45 „ „ „ „ „ 5·24 „
 „ 32 „ 40 „ „ „ „ „ 4·7 „
One man with clothes and other articles . . . . 0·11 „
Provisions with tare and coals for cooking for one man for ·
   one month . . . . 0·07 „
Water with tare for one man for one month . . . 0·14 „
Steam-engines with boilers, water in boilers, coal-boxes, and
   stores per nominal horse-power . . . . . . 0·71 „
Coals for one nominal horse-power in 24 hours . , . 0·15 „

Ballast for sailing-vessels as follows :

For brigs and sloops two and a half the weight of the guns.

For frigates twice the weight of the guns.

For line-of-battle ships one and three-quarters the weight of the guns.

For men-of-war with steam power half the weight of the engines with boilers and water in boilers, or more may be deducted from the ballast as fixed above.

Ships with full steam power do not require ballast for the sake of stability, but they may require a small quantity for trimming the ship.

For line-of-battle ships and frigates the complement of men may be calculated as follows :—

For each 68-pounder of 95 cwt. as a pivot gun . 20 men.
"   "   as a broadside gun . 16 ,.
"   8-inch gun of 65 cwt.   "   " . 11 ,,
"   32-pounder of 56 cwt.   "   " . 10 ,,
"   32   ,,   45 "   "   ,. . 8 ,.
,,   32   ,,   40 "   ,,   ,, . 7 ,,

For the engine-room when the nominal horse-power is $= H$

$$0\cdot2041\ H^{0\cdot8\cdot}$$

The number of men calculated in this way includes all the officers and petty officers.

The number of men may also be calculated as follows :—

To serve the guns one man for every 10 cwt. of the whole number of guns.

For the engine-room and coal-boxes as above : $0\cdot2041\ H^{0\cdot8\cdot}$

For all other service one man for every 7 to 8 cwt. of the guns.

The number of men for merchant vessels may be as follows :—

Cargo in Tons.	Number of Men.	Cargo in Tons.	Number of Men.
15	3	300	13
20	4	400	16
30	5	500	20
40	5	600	22
50	6	800	23
100	8	1000	32
150	9	1500	45
200	11	2000	60

§ 2. On the Qualities of Ships.

a. Merchant Ships.—Merchant ships are designed to carry goods from one place to another across the sea. Their size as well as their qualities must therefore be different, partly according to the merchandise they are to carry, and partly according to the nature of the seas in which they are to navigate ; but generally a merchant vessel should have the following qualities :

1. To sail well, especially by the wind, in order to be able to beat off a coast where it may be embayed, and also to come about well in a hollow sea.

2. To be able to sail with a small quantity of ballast.

3. To work with a crew small in number in proportion to its cargo.

The two first qualities can be obtained by giving the ship great length and breadth in proportion to its displacement and a small depth to the bilge in proportion to the breadth ; but such a vessel requires large sails and heavy anchors, and consequently a great complement of men. The third quality can only be attained by making the ship short, narrow and deep, in proportion to its displacement; but a ship constructed in this manner may not be able to move through the water with the necessary velocity, neither close-hauled nor with the wind abaft the beam. It may, therefore, happen to come into danger, and perhaps be lost for want of good sailing qualities, although it will carry a great cargo in proportion to the expense for the crew and for wear and tear.

As the third condition is in a complete opposition to the two first conditions, and as the ship *must* be able to sail with the necessary degree of safety across the seas, it follows that a merchant vessel, to be as profitable as possible to its owner, cannot possess the very best sailing qualities obtainable.

The vessel that will carry the heaviest cargo with the least expense, and still with a sufficient degree of safety from one place to another, will answer the purpose best. Theoretically to determine the best proportion between the displacement, length, breadth, and depth, is therefore impossible; but the proportions adopted after a practice of more than a couple of thousand years will be the best guide. It is found that the length of sailing vessels varies between three and four times the breadth: in later times, in order to gain a greater velocity the length has been made as much as six times the breadth, but it is doubtful if these very long ships will answer except under particular circumstances, and if not, this proportion will not be generally adopted. The depth from the load water-line to the rabbat of the keel varies between one-half and one-third of the breadth. The larger the ships are, the longer and deeper they are in proportion to the breadth; and this is theoretically correct, though it is certain enough that practice alone has given these proportions to the ships. A proportion between length, breadth, and depth, that in all cases will be better than every other proportion is not to be found, and we also learn from practice that two ships of the same displacement may be very different in regard to the proportions between length, breadth, and depth, and still both be good sea-boats and swift-sailing vessels. The proportions between these three chief dimensions are therefore not of very great consequence when they are not carried to extremes, and it may only be alleged as deduced from practice :

1. That the length ought to be greater in proportion to the breadth in large than in small vessels.

2. That the breadth may be between one-third and one-fourth of the length, and in some particular cases as little as one-sixth of the length.

3. And the depth from the load water-line to the rabbat of the keel from one-half of the breadth in the largest ships to one-third of the breadth in the smallest.

Of more consequence than these proportions, when they are not carried to extremes, is more or less fulness of bottom, having more influence on the ship's qualities, both in regard to sailing qualities and tonnage in proportion to the number of the crew.

If very good sailing qualities are required the ship's bottom must be sharp, with fine lines forward and aft. In this case the hull will be heavy in proportion to the weight it will be able to carry ; and to get the displacement that is necessary for the intended cargo, the ship must be given a greater length, breadth, and depth than needed in case of the bottom being more full or flat, whereby she, of course, will be more expensive to build,

will require larger sails and anchors, consequently also
a greater complement of men, which altogether are
circumstances unfavourable to the owner. If the
bottom is made more flat or fuller towards the
extremities she will, with less length, breadth, and
depth, carry the same load, she will be less expensive
to build, require a smaller number of men to work her,
and be more profitable to her owner ; but she will lose
something of her good sailing qualities. By making her
too full she may lose so much of her sailing qualities
that she will be unable to sail with safety except in fine
weather. It will therefore generally be advantageous
for the owner to be satisfied with a good deal less than
the best sailing qualities that may be obtained by a
vessel of a given displacement.

In determining the fulness of the bottom, the nature
of the seas in which the ship is to navigate, as well as the
nature of the cargo that she generally is to carry, must
be taken into consideration; for a ship designed to
carry merchandise of great value that will bear a heavy
freight, or perishable goods, as, for instance, fruit, may
with advantage be made sharper than a vessel designed
to carry goods of small value in proportion to weight
and volume, such as, for instance, timber and deal.
Likewise vessels that often shall have to cruise
against the trade-winds must have better sailing
qualities than are necessary for other ships, that
are not commonly exposed to encounter such
impediments.

To determine the fulness of the ship's body, we

must resort to experience, although this does not give
any other rules than that the larger ships may be
made more full than the small ones designed for the
same trade, and still retain sufficiently good sailing
qualities. In what manner the fulness of the bottom
may be determined before the drawing is made up will
be shown in the following pages ; but as no fixed rules
can be given for determining the quantity that influ-
ences the fulness of the ship's bottom, it must be
left entirely to the constructor's own experience and
judgment to determine this quantity.

Merchant ships provided with steam power may be
made longer and narrower than sailing vessels, because
their manœuvres do not in narrow seas depend on the
sails, and because they by this measure will obtain
greater speed for the steam power alone.

Steam-ships intended chiefly for passengers are
generally built still more long in proportion to their
breadth, but for sea-going vessels it is probably most
advantageous never to let the length exceed eight times
the breadth.

b. Men-of-War.—By the building of men-of-war the
expenses are only a secondary consideration, in so far
as it will always be the best plan to make them the
most effective fighting ships, the best sea-boats, and
the swiftest-going vessels that circumstances will
admit. As it is impossible to make them to carry
coals for a long time, they must be able to sail at least
tolerably well, and therefore they cannot be without

sails of a certain magnitude; and as they are to carry
guns, generally large ones, their breadth cannot be
limited to that extent as in many other steam-ships.
From this it follows that their speed for steam alone
hardly will be brought to the extent attainable in
steam-ships intended only to carry passengers on given
predetermined and short distances.

In men-of-war, with the greatest steam power until
this time procured for such ships, the length has not
been made more than six times the breadth, and
generally the length is found to be between four and
five times the breadth. Here, as with merchant ships,
it is experience alone that at last will fix the proportion between length and breadth; but it seems not
likely that the length of men-of-war, as sea-going
vessels, at any time with advantage will be made
greater in proportion to the breadth than already now
practised in the longest ships. The depth from the
load water-line to the rabbat of the keel is, in sea-going
men-of-war, found to be a little less than four-tenths of
the main breadth in small vessels, and a little more
than four-tenths of the main breadth in the largest
ships; and this, with so very few exceptions, that there
is every reason to believe that this proportion by experience is found to be the best. Nevertheless, it may be
remarked, that for small vessels, when a light draught
of water is desired, the depth has often been made less
(about one-third of the breadth) without any serious
loss in the good qualities of the ship. The smaller
depth will affect her ability to hold a good wind under

sail, but a small vessel is often built expressly to sail where there is shallow water, and then the light draught of water is of more consequence than the weatherly qualities.

CHAPTER III.

From the Displacement to find the Length, Breadth, Depth, and other Elements of Construction.

To calculate the elements of construction, when the displacement is determined, may be done in different ways; but the most simple that will ensure good qualities to the ship ought naturally to be preferred, and the following method is probably the best hitherto invented. It was invented by the Swedish Admiral, Chapman, and by him called the parabolic system of constructing ships. It rests on the following facts:—

If the areas of the vertical transverse sections of a ship's body below the load water-line are calculated and divided by the greatest breadth of the load water-line, and the quotients according to scale applied on the drawing of the ship on their respective stations from the load water-line downwards, there will be got points through which a fair curved line may be drawn. This curved line is called the curve of sections. This curve of sections will be found to be convex against the water line at both ends; but if it is allowed to run fair from the middle of the ship until it intersects the water-line, these intersections will always be found to be a small distance inside the rabbats of the stem and stern-post.

This distance is generally rather greater forward than aft. The curve of sections constructed in this manner is very often found to agree tolerably well, and in many instances almost exactly with a parabolic line whose exponent is $y^n = px$, and whose apex lies in the greatest section or midship-section. A great number of drawings of ships already built, and whose qualities have been known, have been examined, and it is thereby ascertained that the ships in which the line of sections agreed most accurately with a parabolic line invariably bore excellent characters.

The exponent of this curve is found to vary from 1·8 to 2·8, according to the fineness of the ship's ends in proportion to the midship-section, and is generally found to be greater in large ships than in small ones. The parabolic system of construction deduced from these facts cannot theoretically be proved better than other systems; but on the following reasons it is thought preferable to all other methods until this time published :—

1. It may be said to be derived from experiments made with ships actually built and of different descriptions, and therefore it ensures that the ship with a certain fulness of bottom shall obtain the best possible qualities in regard to good sailing and easiness in a sea-way.

2. It is possible by this system of construction, without any great trouble, to make up a drawing for a ship exactly to a predetermined displacement, and in a manner that the centre of gravity of displacement

shall come exactly in the place according to the ship's length that is thought most favourable for good steer· ing, manœuvring, and stowage, which is of great conseqnence.

3. It is easy of application, and gives every desirable opportunity to vary the form of the ship's body according to the views of the constructor.

Although the expouent of the curve of sections varies considerably in different ships, it may, for all sailing merchantmen, be taken to be 2·5, and only when a ship is to be very large, a greater exponent, for instance 2·6, can be recommended. If the principal point is to obtain more than ordinary good sailing qualities, the exponent of the line of section ought to be taken less than 2·5, say from 2 to 2·3, according to the size of the ship.

For men-of-war and steamers it may be laid down as a general rule that the exponent of the line of sections is to be from 2 to 2·3, so that for a displace· ment of 5000 cubic feet or less the exponent is to be 2 or 2·1 ; for a displacement from 5000 to 10,000 cubic feet the exponent is to be 2·1 or 2·2; and for a displace· ment greater than 10,000 cubic feet, the exponent of the line of sections may be 2·2 or 2·3.

The load water-line or first water-line is that line to which the ship with propriety may be loaded, and its length is calculated from the inside of the rabbat of the stem to the inside of the rabbat of the stern·post. Through these points on the drawing are drawn lines perpendicular to the water-line, and therefrom the

c

expression, "length between the perpendiculars." Not
all the constructors of ships draw the perpendiculars
through these points, but in this book it is always by
this expression understood the length of the load
water-line from the inside of the rahhat of the stem to
the inside of the rahhat of the stern-post.

As before mentioned, the parabola that coincides
for the greatest length with the line of sections,
intersects the load water-line a little on the inside
of the perpendiculars, and the length of that part
of the water-line which lies between the two points
in which the parabola intersects the load water-line,
is called the length of the construction water-line. It
is found that the difference between the length of the
construction water-line and the length between the
perpendiculars for all sailing ships, and vessels of all
sizes, may he put=2·5 feet, of which one foot is added
to the length of the construction water-line abaft and
1·5 feet forward. For men-of-war and steamers this
difference ought to be greater, say from 4 to 10 feet;
and particularly in screw steamers it will be found
necessary to make the addition to the length of the
construction water-line greater abaft, to obtain free
access of the water to the screw propeller. How large
this addition shall be will depend upon the size of the
propeller.

When the displacement is found according to the
rules given in Chapter II., it is the number of cubic
feet contained in that part of the ship's body which is
below the load water-line, the planking, stem, stern-

post, and keel not included, and is therefore called the displacement on the timbers. Put this displacement $= D$, the exponent of the said parabola or curve of sections $= n$, the length between the perpendiculars $= L$, and the length of the construction water-line $= l$. Then is, for sailing-vessels, $L = l + 2.5$, and for men-of-war and steamers, $L = l +$ a certain distance. Put the breadth on the timbers in the load water-line at the greatest transverse section, or what is called the midship section $= B$, and the abscissa of the curve of sections on the midship section, that is, the area of the midship section divided by $B, = h$. Then the area of the curve of sections is $= \dfrac{n}{n+1} lh$, the displacement $= D$

$= \dfrac{n}{n+1} lBh$, and the area of the midship section is $= Bh$. Put the depth from the load water-line to the upper edge of the rabbat of the keel on the midship section $= d$, and then the area of the midship section may be put $= mnd$; where m is always less than unity, and shows the proportion between the area of the mid-ship section and that of the circumscribed parallelo-gram, in which one side is $= B$, and the other $= d$.

As before-mentioned, the breadth is not in all sailing vessels to be in the same proportion to the length, and therefore the breadth $= B$ may be put $= pl^r$. In a like manner the depth d may be put $= kB^r$, by which d is found $= kp^r l^{rr}$, and if kp^r is put $= v$ and $vr = x$, d is $= ol^x$.

The area of the midship section $= \phi$ is $= Bh = mBd$, and when the values of B and d, viz. pl^r and ol^x, are substituted, ϕ is $= \text{B}h = m\,o\,p\,l^{v+r}$. This equation multiplied by $\dfrac{n}{n+1}l$, gives $\dfrac{n}{n+1}l\text{B}h$, or $\text{D} = \dfrac{n}{n+1}mopl^{v+x+1}$. (A.)

By comparison of a great number of ships it is found that for all sailing merchantmen v may be $=0.8$ and $p = 0.7$, $x = 1$ and $o = \frac{1}{4}$, by which B will be $= 0.7l^{r\cdot 4}$ and $d = \frac{1}{8}l$. $v + x + 1 = 0.8 + 1 + 1 = 2.8, \dfrac{n}{n+1}op = \frac{2\cdot 5}{3\cdot 5} \times \frac{1}{4} \times 0.7 = \frac{1}{10}$.

When the values of $\dfrac{n}{n+1}op$ and of $v + x + 1$ are substituted in the equation A, it gives $\text{D} = m \times \frac{1}{10}l^{2\cdot 8}$, and therefore $\dfrac{1}{m} \times 16 \times \text{D} = l^{2\cdot 8}$. The length of the construction water-line is then found $= l = \left(\dfrac{1}{m}\right)^{\frac{10}{28}} \times 16^{\frac{10}{28}} \times \text{D}^{\frac{10}{28}}$
$= \left(\dfrac{1}{m}\right)^{\frac{10}{28}} \times 2.6918 \times \text{D}^{\frac{10}{28}}$.

By this equation we have the length of the construction water-line when m is determined, and then the following dimensions, viz :—

> The breadth $= \text{B} = 0.7l^{r\cdot 4}$.
> The depth $= d = \frac{1}{8}l$.

The length between the perpendiculars $= \text{L} = l + 2.5$.
The area of the midship section $= \phi = \text{B}h = m\text{B}d$.
The abscissa of the curve of sections on the midship section $= h = \dfrac{\phi}{\text{B}} = md$.

It is now seen that the coefficient m must be determined before any other dimensions can be calculated, and that this same coefficient alone regulates the ful-

ness of the bottom for all merchant ships not intended
to have steam power. The said coefficient may vary
between 0·5 and 1·0, but it seldom reaches these extre·
mities. In ships designed for the same trade it may be
greater for the large than for the small ones, so that
for a common merchantman, whose displacement is from
25,000 to 30,000 cubic feet, m may properly be from
0·84 to 0·86; while in a merchantman whose displacement is only 4000 to 5000 cubic feet, it ought not to
be larger than 0·75. The constructor must determine
upon this quantity according to his own judgment
and experience.

For men-of-war and other steam-ships the calculated
displacement is as before $= D = \dfrac{n}{n+1} mlBd$ (B).

We may here put $B = \dfrac{l}{p}$ and $d = kB = \dfrac{kl}{p}$. By substi-
tuting these values of $B \overset{l}{\sqrt{}}$ and d for $B, \overset{l}{\sqrt{}}$ and d in the
equation (B) we have $D = \dfrac{n \, m \, k}{(n+1) \, p^{3}} \, l^{3}$ or $D = \dfrac{n \, m \, p \, k}{n+1} B^{3}$ and

consequently $l = \sqrt[3]{\dfrac{(n+1) \, p^{2} \, D}{n \, m \, k}}$ and $B = \sqrt[3]{\dfrac{(n+1) \, D.}{n \, m \, p \, k}}$

By fixing n, m, p and k, according to circumstances,
l or B may be first found, and thereafter d and ϕ, or the
area of the midship section.

For three-decked ships p may be from 4 to 4·5 and k from 0·41 to 0·43
For line-of-battle ships p „   „ 4·5 „ 5   „ k „ 0 4 „ 0 42
. For frigates   . . p „   „ 4 5 „ 5   ., k „ 0 4 „ 0·42
For flush-decked ships p „   „ 5 „ 6 5 „ k „ 0·37 „ 0·4

For steam-boats, intended for goods and passengers
only, p may be greater and k less than for men-of-war ;

and for river boats p is sometimes made as great as 10, and k as small as $0 \cdot 1$.

When the length of a ship or vessel is more than three and a half or four times the breadth, it is generally advantageous to make a part of her in the middle of the length such that the greatest transverse section or midship section is unaltered for the whole length of this part. As the sides of the ship in this part are parallel, I will call this part a parallel piece. If l is $=pB$ as before, and p greater than 4, the length of the parallel piece may be $= l - 4B = pB - 4B = (p-4)$ B. The displacement of the parallel piece $= D''$ is then $= (p-4)$ Bϕ. The displacement of the fore and after part $= D'$ will be $\dfrac{n}{n+1}4B\phi$, and the whole displacement $- D = D'$

$$+ D'' = \left(\frac{4n}{n+1} + p - 4\right)B\phi = \left(\frac{4n}{n+1} + p - 4\right)mkB^3, \text{ and } B =$$

$$\sqrt[3]{\frac{D}{\left(\frac{4n}{n+1} + p - 4\right)mk}};$$ the letters signifying the same

as above. This process will, with the same n, m, p and k, give a smaller and consequently lighter hull, and a smaller midship section, which, as the resistance of the water against the ship's body chiefly depends on the area of this section, very likely will produce a greater speed for steam alone; at the same time lessening the building expenses, and, as far as my experience goes, ensuring easier motions in a rough sea.

After the length, breadth, and depth are calculated, the place of the midship section in regard to the length must be determined upon. As this depends upon where it is thought proper to have the centre of gravity of the displacement, it is necessary that the place of this centre with regard to the length is fixed. For the management of sailing-ships it would be the best to have the centre of gravity in the middle of the length between the perpendiculars, but as the anchors, bowsprit, and foremast, with their spars and rigging, are considerable weights in the forepart of the ship, and the cabin, &c., take away a considerable room abaft from the part of the ship that can be loaded, it is impossible so to construct a ship that the centre of gravity shall come in the middle of the length, without making it necessary to take on board a quantity of ballast, only to counterbalance the weight of the anchors, bowsprit, and fore-rigging. Such ballast is a dead weight, taking away part of the hold, that should be used exclusively for the lading; and as it is of consequence in a merchantman, into a ship of a given displacement, to be able to stow the greatest possible quantity of goods, it is found more advantageous to construct the ship so that the centre of gravity shall come a small distance before the middle of the length when the ship is loaded to the load water-line. By this means all ballast, except what with certain descriptions of cargo may be necessary to ensure stability, is dispensed with.

In ships built at different times and at different places, the distance between the vertical through the centre of gravity of the displacement and the middle of the length of the load water-line is found to vary considerably. In some ships the centre of gravity is found to be about $\frac{1}{70}$ of the length of the load water-line, in others about $\frac{1}{80}$ of the same length before the middle of the same water-line; but, deduced from the best ships, the centre of gravity of the displacement, in all sailing merchant vessels, may properly be placed $\frac{1}{70}$ of the length between the perpendiculars, that is $\frac{1}{70}$ L, before the middle of this same length. Now, in this class of ships the middle of the length between the perpendiculars is 0·25 feet before the middle of the length of the construction water-line (see the foregoing); and consequently the centre of gravity before the vertical passing through the middle of the construction water-line $\frac{1}{70}$ L + 0·25. If n is the exponent of the line of sections, the midship section will come before the centre of gravity $\left(\frac{1}{70} L + 0·25\right)(n+1)$ and before the middle of the construction water-line $\frac{1}{70} L + 0·25 + \left(\frac{1}{70} L + 0·25\right)(n+1) = \left(\frac{1}{70} L + 0·25\right)(n+2)$.

It is proved by experience, to a tolerable degree of certainty, that there is a sharpness of the after-body more advantageous for speed than any other, but also that a small deviation from this sharpness does not perceptibly affect the velocity. By the common proportion between length and breadth in sailing-vessels, this sharpness will be obtained very nearly by following the rules given above; and with a different proportion between length and breadth, but with the common proportion between breadth and depth, the probably most advantageous sharpness of the after-body will be acquired by placing the midship section about twice

the main breadth before the after-end of the construction water-line. If the depth is less than general
in proportion to the breadth, the midship section comes
nearer the after-end of the construction water-line;
sometimes so near as 1·5 times the main breadth.
Likewise it is certain that with a given area of midship
section, the resistance of the fore-body will be diminished in some proportion to the fineness of the
lines forward. When, therefore, the length is more
than four times the breadth, the resistance of the water
will be less when the greatest transverse section is
placed abaft the middle of the load water-line, and
consequently also the centre of gravity abaft the middle
of the same water-line. The centre of gravity may,
however, be brought farther forward by giving the ship
a parallel piece in the middle; and this proceeding has
also other advantages, as before mentioned.

If the length of the construction water-line is l, the
length of the parallel piece $\dfrac{l}{v}$, the distance that the
centre of gravity shall be before the middle of the
construction water-line z, and the exponent of the line
of sections for the fore and after-body n, the middle
of the parallel piece will be before the centre of gravity
$\dfrac{vn(n+1)+n}{vn+2} z$, and the middle of the parallel piece
before the middle of the construction water-line
$\dfrac{(vn+1)(n+2)}{vn+2} z$. If v is infinitely great, that is,
$\dfrac{l}{v} = o$, there is no parallel piece, and we have $(n+1) z$

and $(n+2)$ z, as above (z being $= \frac{1}{30}$ L $+0.25$ for merchant vessels).

The displacement is $= $ D $= \frac{l}{v}\,\phi + \frac{n}{n+1}\left(l-\frac{l}{v}\right)\phi = \frac{vn+1}{v(n+1)}\,l\,\phi$ and $\phi = \frac{\mathrm{D}\,v\,(n+1)}{l\,(vn+1)}$.

In steam-ships, not intended to use sails except as an auxiliary power, it is not of any great consequence to get the centre of gravity in a certain place with regard to the length of the ship, and therefore the greatest transverse section may be placed about twice the main breadth from the after-end of the construc-tion water-line, and the distance z of the centre of gravity from the middle of the length of the construc-tion water-line may be found by the formula above.

In order that a sailing-vessel shall not be too weatherly or ardent, or what amounts to the same, not require too large fore-sails, it is necessary to construct her to draw more water abaft than forward. Experience tells us that this difference of draught of water pro-portionally must be greater in a small vessel than in a large one, and therefore it may be as well to make the difference of draught of water forward and aft the same for all sailing-vessels, viz., 1·25 feet. When the distance between the midship section and the middle of the length between the perpendiculars is c, and the difference of draught of water is k, the distance from the load water-line to the upper edge of the rabbat of the keel on the aftermost perpendicular is $= d + \frac{k}{2}$

$+\dfrac{k\,c}{\text{L}}$, and on the foremost perpendicular $= d -$
$\dfrac{k}{2}+\dfrac{k\,c}{\text{L}}$, L being the length between the perpendiculars, and d the depth from the load water-line to
the upper edge of the rabbat of the keel on the
midship section.

The rake of the stern-post from the load water-line
to the rabbat of the keel may, on vessels of all sizes,
be 1 foot, and the rake of the stem from the load water-
line to the rabbat of the keel equal to one-fourth of
the main breadth $=\tfrac{1}{4}$B. Some constructors think it
better to give the stem less rake : it seems to have
very little influence on the ship's qualities, and may be
considered as a matter of taste.

The fore part of the water-line from the midship
section to the stem may, in sailing-vessels, be made a
parabolic line, whose exponent is $1\cdot51\sqrt{\text{\textasciitilde}}$, where h is
the abscissa of the line of sections on the midship
section $=\dfrac{\phi}{\text{B}}$.

The breadth, depth, and area of the midship section
is already found, but it will easily be understood that
with these calculated elements the midship section may
still be given very different forms, and the limits for
these, with a given area, breadth and depth, may be
seen (Pl. I., fig. 17) by the two figures $a\,b\,c\,d$ and
$a\,b\,e\,f\,g$, both of which have the same area, breadth,
and depth.   The stability of the ship, considered without regard to the stowage of the cargo, depends chiefly

on the form of the body below the load water-line;
and as the form of the midship section to a certain
degree regulates the forms of all the other transverse
sections, it is of great consequence in sailing-ships to
give the midship section the shape that will contribute
the most to the stability of the ship, that the sails may
be made sufficiently large to give the ship good sailing
qualities. When the cargo is stowed on judicious
principles, it is not to be feared that the ship shall be
too stiff, although it may have heavy goods on board,
as for instance salt or iron; and with light goods on
board, the stability will not be greater than necessary,
when the proportions hereafter given are applied to the
construction of the midship section. In steam-ships
that are given small sails in proportion to the dimensions of the hull, it may be advisable to give the ship's
body below the load water-line a form that will make
the ship less stiff, in order that her motions in a rough
sea shall not be uneasy. ·

In the figure $a\,b\,c\,f\,g$ (Pl. I., fig. 15), the centre of
gravity is considerably nearer the line ab, that is to
represent the water-line, than in the figure $a\,b\,c\,d$, and
consequently the figure $a\,b\,c\,f\,g$ will give the greatest
stability. In sailing-ships, therefore, the form of the
midship section ought to partake as much as possible
of this form. The breadth of the water-line is the
chief dimension on which the stability depends; and
that a ship may not lose too much of her stability
when obliged to sail with less than full cargo, or with
ballast alone, the main breadth ought to be continued

a distance below the load water-line, and in this respect also the figure $a\,b\,e\,f\,g$ is superior to others. Still it follows that the centre of gravity of the cargo will come a little higher when the figure $a\,b\,e\,f\,g$ is adopted for the midship section than when the figure $a\,b\,c\,d$ is used, and by this circumstance a little of the stability gained by the adopted form of the midship section will be lost; but this loss is not considerable; so that, of two ships of the same principal dimensions, displacement, and area of midship section, but with different forms of this section, the one whose midship section coincides the nearest with the figure $a\,b\,e\,f\,g$, is decidedly the stiffest.

The height from the load water-line to the deck at the lowest part of the ship may, with propriety in common merchant ships, be $= 0.04444\, \mathrm{B}^{1.4}$, where B is the main breadth as before. In clippers, this height may be a little less.

In men-of-war, the height from the load water-line to the lower portsill may he: for three-decked ships $= \dfrac{\mathrm{B}}{9}$, for line-of-battle ships $= \dfrac{\mathrm{B}}{7.5}$, for frigates $= \dfrac{\mathrm{B}}{5.6}$ to $\dfrac{\mathrm{B}}{5.8}$, and for sloops, or flush-decked ships $= \dfrac{\mathrm{B}}{4.4}$ to $\dfrac{\mathrm{B}}{5}$. In small steamers intended for passengers, the height from the load water-line to the deck may be $= \dfrac{\mathrm{B}}{3.6}$ to $\dfrac{\mathrm{B}}{4}$.

When, in the general equation of a parabolic line $y^n = px$, the exponent n is given the values $1 - 1.1 -$

1·2, and so on to 6, y the values from 1 to 10, and p always such a value that x is = 1 when y is = 10; the different values of x are found as put down in the following table:—

Exponent = 1·	1·1	1·2	1 3	1·4	1 5	1·6	1 7	1 3	1·9	2·
y x	x	x	x	x	x	x	x	x	x	x
1 0·1	0·0794	0·0631	0·0501	0·0398	0·0316	0·0251	0·0179	0·0158	0·0126	0·01
2 0·2	0·1700	0·1451	0·1233	0·1051	0·0895	0·0761	0·0648	0·0551	0·0470	0·04
3 0·3	0 2668	0·2360	0·2090	0·1855	0 1644	0·1457	0·1289	0·1146	0·1015	0·09
4 0·4	0·7654	0·3331	0·3038	0·2771	0·2530	0·2308	0·2107	0·1922	0·1754	0·16
5 0·5	0·4662	0·4355	0·4060	0·3790	0·3536	0·3298	0·3078	0 2872	0·2685	0·25
6 0·6	0·5703	0·5420	0·5148	0·4895	0·4649	0·4418	0·4196	0·3987	0·3788	0·36
7 0·7	0·6751	0·6517	0·6290	0·6071	0·5857	0·5652	0·5453	0·5261	0·5077	0·49
8 0·8	0·7824	0·7654	0·7484	0·7317	0·7157	0·7000	0·6844	0·6691	0·6544	0·64
9 0·9	0·8904	0·8807	0·8722	0·8627	0·8539	0 8448	0·8350	0·8273	0·8186	0·81
10 1·	1·	1·	1·	1·	1·	1·	1·	1·	1·	1·

Exponent 2·1	2 2	2·3	2·4	2 5	2 6	3·	3 5	4	4 5	5·	6·
y x	x	x	x	x	x	x	x	x	x	x	x
1 0·0079	0·0063	0·0050	0·0040	0 0032	0·0025	0·001	0·0003	0·0001	0·0000	0·0000	0·0000
2 0·0341	0·0290	0·0246	0·0210	0·0178	0·0152	0·008	0·0036	0·0016	0·0007	0·0003	0·0001
3 0·0795	0·0707	0·0627	0·0556	0·0493	0·0437	0·027	0·0148	0·0081	0·0044	0·0024	0·0007
4 0·1460	0·1332	0·1215	0·1109	0·1010	0·0923	0·064	0·0405	0·0256	0·0162	0·0102	0·0041
5 0·2333	0·2176	0·2031	0·1894	0·1768	0·1650	0·125	0·0884	0·0625	0·0442	0·0313	0·0156
6 0·3421	0·3250	0·3088	0 2935	0·2788	0·2650	0·216	0·1673	0·1296	0·1002	0·0778	0·0467
7 0·4723	0·4562	0 4403	0·4248	0·4100	0·3956	0·343	0·2870	0·2401	0·2008	0·1681	0·1176
8 0·6259	0·6121	0 5988	0·5853	0·5724	0·5598	0·512	0·4579	0·4096	0·3664	0·3277	0·2621
9 0 8015	0·7931	0·7843	0·7765	0·7684	0·7604	0·729	0 6916	0·6561	0·6224	0·5905	0·5314
10 1·	1·	1·	1·	1·	1·	1·	1·	1·	1·	1·	1·

By help of this table the figure 16, Plate II., is con-structed in the following manner. The line A B is, according to scale divided after the exponents, so that A1·5 is = 1·5, A2 is = 2, A2·5, is = 2·5 and so on. The lines A C and B D are drawn perpendicular to A B, and each of them made equal to unity, according to a scale on which may be measured thousand parts. Join C and D with a straight line and draw the lines through the divisions, 1·5, 2, 2·5, &c. parallel to A C and B D. On these parallel lines apply the ordinates x in the table, the ordinates for the exponent 1·5 on the line 1·5, the ordinates for the exponent 2 on the line 2, and so on from A B downwards, and draw then the curved lines 1, 2, 3, 4, 5, &c. through the marks for the ordinates of the same number. The ordinates are to be taken on the scale to which A C was made equal unity. By these figures it will be easy to draw parabolas with exponents from 1 to 6, as afterwards will be shown.

CHAPTER IV.

To calculate the Elements of Construction of Ships.

SUPPOSE a ship is to be built to carry a load of 500 tons for common trading purposes, the quantity m showing the proportion between the area of the midship section and that of the circumscribed parallelogram, may then be $= 0.84$, and according to Chapter II., sect. 1, the displacement is $= L + \dfrac{0.5}{m} L = 500$

$+ \dfrac{0.5}{0.84} \times 500 = 500 + 297.62 = 797.62$ tons, which, multiplied by 35, gives the displacement $D = 27916.7$ cubic feet.

The length of the construction water-line, according to Chapter III., is $l = \left(\dfrac{1}{0.84}\right)^{\frac{10}{4 \cdot k}} \times 2.6918 \times D^{\frac{1}{3}}$ and the principal dimensions are found by the following calculations :—

$$
\begin{aligned}
\text{Log } 1 &= 10.00000 - 10 \\
\text{Log } 0.84 &= 9.92428 - 10 \\
\hline
\text{Log } \tfrac{1}{0.84} &= 7)0.07572 \\
\hline
& 0.4)0.010817 \\
\hline
\text{Log } \left(\tfrac{1}{0.84}\right)^{\frac{10}{4k}} &= 0.02704
\end{aligned}
$$

$$\text{Log D} = {}^{i})4\cdot44587$$
$${}^{\infty 4})0\ 635124$$
$$\text{Log D}\tfrac{10}{13} = \underline{\quad 1\cdot58781}$$
$$\text{Log}(\tfrac{1}{0\cdot64})\tfrac{10}{23} = 0\cdot02704$$
$$\text{Log } 2\cdot6918 = 0\cdot43004$$
$$\text{Log } l = 2\cdot04489 \text{ and } l = 110\cdot89$$

0 8   2·5 = the difference

$$\text{Log } l^{0\,8} = 1\cdot635912 \qquad \text{between } l \text{ and L.}$$
$$\text{Log } 0\cdot7 = 9\cdot84510 - 10 \quad 113\cdot39 = \text{L} = \text{the length.}$$
$$\text{Log } 0\ 7\ l^{0\,8} = \text{Log B} = 1\ 48101 \text{ and } \text{B} = 30\cdot27 = \text{the breadth.}$$

The depth $d = \tfrac{1}{8}\, l = 13\cdot861$ and $\text{Log } d = 1\cdot14179$

$$\text{Log } m = 9\cdot92423 - 10$$
$$\text{Log } md = \text{Log } \lambda = 1\cdot06607 \text{ and } \lambda = 11\cdot643 = \tfrac{\phi}{\text{B}}$$
$$\text{Log } \text{B} = 1\cdot48101$$
$$\text{Log } \phi = 2\cdot54708 \text{ and } \phi = 352\cdot43 \text{ square ft.}$$

$$\text{Log } \lambda = {}^{7})1\cdot06607$$
$$\text{Log } \sqrt{\lambda} = 0\cdot53303$$
$$\text{Log } 1\cdot51 = 0\cdot17898$$

$\text{Log } n^1 = 0\cdot71201$ and $n^1 = 5\cdot152 =$ the exponent of the forepart of the load water-line.

$$\text{Log B} = 1\cdot48101$$
$$1\cdot43$$
$$\text{Log } \text{B}^{1\,73} = 2\cdot11784$$
$$\text{Log } 0\cdot0444 = 8\cdot64777 - 10$$

$\text{Log } g = 0\ 76561$ and $g = 5\cdot83 =$ the height from the load water-line to the deck.

The centre of gravity of the displacement before the middle of the load water-line $= \dfrac{\text{L}}{50} = 2\cdot2678$:

* The same centre before the middle of the construction water-line

$\dfrac{\text{L}}{50} + 0\cdot25 = 2\cdot5178 = x$;

ϕ before the middle of the construction water-line $= (n + 2)\ x = 4\cdot5 \times 2\cdot5178 = 11\cdot3301$:

ϕ from the fore-end of the construction water-line .   44·115
Do.   „   after end of   do.   .   .   .   .   66·775
Do.   „   foremost perpendicular, 44·115 + 1·5 =   .   45·615
Do.   „   aftermost   do.   66·775 + 1 =   .   67·775
Do. before the middle of the load water-line, 11·33 − 0·25 =
    11·08 = c.

Difference of draught of water = k = 1·25.

Depth from the load water-line to the upper edge of the rabbat of the keel on the aftermost perpendicular $= d + \dfrac{k}{2} + \dfrac{k\,c}{\mathrm{L}} = 14·608.$

Do. on the foremost   do. $= d - \dfrac{k}{2} + \dfrac{k\,c}{\mathrm{L}} = 13·358.$

Rake of stem from the rabbat of the keel to the load water-line = $\tfrac{1}{3}\mathrm{D}$ = 7·57.

Rake of the stern-post from the rabbat of the keel to the load water-line = 1 foot.

If now such a ship, intended to carry 500 tons, is desired to be more swift-sailing than generally is found profitable for merchantmen, but which, however, may be advantageous under certain circumstances, or is to be what is called a clipper ship, m may be = 0·75 and n = 2·3. The depth d from the load water-line to the rabbat of the keel on the midship section may then be made in the same proportion to the length, that by experience is found to answer well for flush-decked men-of-war without steam power, viz. $= \tfrac{1}{16} l$. We have then in the formula $\mathrm{D} = \dfrac{n}{n+1}\, m\, o\, p\, l^{v+x+1}$ $n = 2·3, m = 0·75, o = \tfrac{1}{16}, p = 0·7, v = 0·8,$ and $x = 1,$ and l is found to be $= \left(\dfrac{33\,\mathrm{D}}{1·2075}\right)^{\frac{14}{44}}.$

The displacement D is then, according to Chapter II.,

§ 1, $\mathrm{L} = + \dfrac{0·6}{m}\,\mathrm{L} = 500 + \dfrac{0·6}{0·75} \times 500 = 500 + 400 =$

900 tons, which, multiplied by 35, gives $D = 31500$ cubic feet.

The calculations will then stand as follows :—

Log D $=$ 4·49831
Log 33 $=$ 1·51851

Log 33 D $=$ 6·01682
Log 1·2075 $=$ 0·08189

Log $\frac{33 P}{1·2075}$ $=$ 5·93493

            0·4)0·847347

Log l $=$ 2·11962 and $l = 131·71$
           0·8   2·5 = the difference between l and L.

Log $l^{0·8}$ $=$ 1·69596   134·21 = L = the length.
Log 0·7 $=$ 9·84510

Log B $=$ 1·54060 and $B = 34·74 =$ the breadth.
The depth $d = \frac{1}{10} l = 13·171$ and $\log d = 1·11962$

           Log m .   . $= 9·87506 - 10$

           Log $md = \log h = 0·99468$ and $h = 9·878$
           Log B .   . $= 1·54080$

           Log ϕ .   . $= 2·53548$ and $\phi = 343·15$ sq. feet.
Log h $=$ 2)0·99468

Log $\sqrt{h}$ $=$ 0·49734
Log 1·51 $=$ 0·17898

Log n^1 $=$ 0·67632 and $n' = 4·746 =$ the exponent of the forepart
      of the load water-line.
Log B $=$ 1·54080
           1·43

Log $B^{1·43}$ $=$ 2·20334
Log 0·0444 $=$ 8·64777 - 10

Log g $=$ 0·85111 and $g = 7·098$, which, as it is a clipper ship,
      may be reduced to 6·5 = the height from the load water-line to
      the deck.

The centre of gravity before the middle of the load water-line $-\frac{L}{50} = 2·6842$, and the same centre before the middle of the construction water-line $\frac{L}{50} + 0·25 = 2·9342 = x$.

φ before the middle of the construction water-line = $(n+2)\, z =$ 4·5 × 2·9342 = 12·617.

φ from the fore-end of the construction water-line		.	53·238
Do.	„ after-end of	do.	78·472
Do.	„ fore-end of the load water-line, 53·238 + 1·5		54·738
Do.	„ after-end of	do. 78·472 + 1	79·472

Do. before the middle of the load water-line = 12·617
  — 0·25 = ... 12367 = c

Difference of draught of water = k = 1·25.

Depth on the aftermost perpendicular, from the load water-line to the upper edge of the rabbat of the keel, $= d + \dfrac{k}{2} + \dfrac{kc}{L} = 13·911$.

Do. on the foremost   do. $= d - \dfrac{k}{2} + \dfrac{kc}{L} = 12·661$.

Rake of stem from the upper edge of the rabbat of the keel to the load water-line $= \frac{1}{4}$ B = 8·685.

Rake of stern-post from the upper edge of the rabbat of the keel to the load water-line = 1 foot.

The resistance of the water against the ship's progress depends chiefly on the area of the midship section, and the midship section of the common merchant ship is to that of the clipper ship as 1 to 0·97. The area of the sails, in case of their being in both vessels proportioned to the dimensions of the vessels in the same manner, is as 1 to 1·365, and in this instance the moment of sails will be as 1 to 1·56; but the stability of the two ships is nearly as 1 to 1·8; consequently the clipper will be able to carry proportionally to her dimensions still greater sails, and shall of course sail better than the common merchantman, to which also her finer lines will assist her considerably. The clipper will, on the contrary, have the disadvantage to require more men to work her, and her register tonnage will be greater, making the running expenses greater, and the cost of building her will be about 14 per cent. more than the cost of the common merchantman. Lastly, she will probably be more uneasy in a sea-way. The clipper might have been made still sharper, and very likely have been able to sail yet faster; but then all the disadvantages above mentioned would have been greater, and probably with heavy goods on board she might have been uneasy and run the risk to spring a leak and damage her cargo.

As further examples we will calculate the elements of construction of a schooner of 94 tons, intended for a swift-sailing merchantman, and consequently sharp as such, and an East Indiaman of 1000 tons.

Let m for the schooner be $= 0.7$, and as this vessel is sharp, the weight of the hull will be $= \dfrac{0.6}{0.7} \times 94 = 80.6$, and the whole displacement $= 174.6$ tons $= 6111$ cubic feet $= D$.

For the East Indiaman let m be $= 0.86$, then the weight of the hull will be $= \dfrac{0.6}{0.86} \times 1000 = 697.7$ tons, and the displacement $= 1697.7$ tons $= 59420$ cubic feet $= D$.

By the formula $l = \left(\dfrac{1}{m}\right)^{\frac{10}{48}} \times 2.6918 \times D^{\frac{10}{44}}$ &c., the elements of construction of these two ships are found as put down in the table that follows.

As an example of a man-of-war, suppose a frigate is to be designed, and that this frigate is to carry—

On the main-deck 32 8-inch guns of 65 cwt. each.
On the upper-deck 18 32-pounders of 45   „
        and 2 68-   „   95   „
                ___
        Sum total 52 guns.

That she shall have a steam-engine of 500 nominal
horse power, carry coals for 6 days' full steaming, provisions for 2 months, and water for 1 month, with a
distilling apparatus.

Thus, according to Chapter II., we have—

32 8-inch guns with ammunition, gunners' stores, and all
   other appurtenances, weighing 32 × 7·2 tons, is   .   230·4 tons.
18 32-pounders with ditto „   18 × 5·24 „ . . 94·32 „
 2 68-   „   „   „   2 × 12·5 „ . . 25·0 „

COMPLEMENT OF MEN.

32 8-inch guns of 65 cwt., 11 men for each   . 352 men.
18 32-pounders of 45 „ 8   „   . . 144 „
 2 68-   „   95 „ 20   „   . . 40 „
For the steam-engine $0·2041 \times 500^{0·6}$.   . . 40 „
                                         ___
     Total number of men 576

Which number multiplied by 0·11 gives   .   .   . 63·36 „
Provisions for 2 months = 576 × 2 × 0·07   .   . . 80·64 „
Water for 1 month = 576 × 0·14 .   .   .   . 80·64 „
Steam-engines of 500 horse power = 500 × 0·71   .   . 355·0 „
Coals for 6 days' steaming = 6 × 500 × 0·15 .   .   450·0 „
In a sailing-frigate with this armature the ballast ought
to be about 300 tons; but as the steam-engines weigh
355 tons, about 200 tons may be deducted, and the
ballast .   .   .   .   100·0 „
                                     ___
                            s = 1479·36 „
Weight of hull = 1·25 s = ... 1849·20 „
Masts, spars, rigging, sails, &c. = 0·2 s =   .   . 295·87 „
                                     ___
     Total displacement = ... 3624·43 tons,

which, multiplied by 35, gives the displacement = D = 126855 cubic
feet.

In a ship of this class the length of the construction water-line may properly be 5·5 times the breadth, and in the formula, Chapter III., $B = \sqrt[3]{\dfrac{D}{\left(\dfrac{4n}{n+1} + p - 4\right) m\,k}}$; p is then $= 5\cdot5$, $n = 2\cdot3$, and, as it is a steam-frigate, m may be $= 0\cdot8$, to give a good room for the engines and boilers; k may be $= 0\cdot4$, and the calculations will stand as follows :—

$$n = 2\cdot3,\ 4n = 9\cdot2$$
$$n + 1 = 3\cdot3\ \angle$$
$$\frac{4n}{n+1} = 2\cdot7879$$
$$p - 4 = 1\,5$$
$$\frac{4n}{n+1} + p - 4 = 4\cdot2879 \quad \log = 0\cdot63224$$
$$mk = 0\cdot8 \times 0\cdot4 = 0\cdot32 \quad \log = 9\cdot50515 - 10$$
$$\mathrm{Log}\left(\frac{4n}{n+1} + p - 4\right)mk = \qquad .\qquad . \quad 1\cdot13739$$

$$\mathrm{Log}\left(\frac{4n}{n+1} + p - 4\right)mk = 1\cdot13739$$
$$\mathrm{Log\ D} = \log 126855 = 5\cdot10331$$
$$3)4\cdot96592$$
$$\mathrm{Log}\ B = 1\cdot65531 \text{ and } B = 45\cdot218$$
$$\mathrm{Log}\ 5\cdot5 = 0\cdot74036$$
$$\mathrm{Log}\ l = 2\cdot39567 \text{ and } l = 248\cdot7$$

The addition to the construction water-line . 9·3

Length between perpendiculars = L = . 258·0

The addition forward may be 3, and abaft 6·3 for the screw.

$$\mathrm{Log}\ B = 1\cdot65531$$
$$\mathrm{Log}\ 0\cdot4 = 9\cdot60206 - 10$$
$$\mathrm{Log}\ d = 1\cdot25737 \text{ and } d = 18\cdot087$$
$$\mathrm{Log}\ m = 9\cdot90309$$

Log h = 1·16046 and h = 14·47
Log b = 1·65531
Log ϕ = 2·81571 and ϕ = 654·3 square feet.
b = 45·218
        4
$4b$ = 180·872
l = 248·7
$l-4b$ = 67·828 = the length of the parallel piece = P.

In this ship the centre of gravity may with propriety be in the middle of the length of the construction water-line, and therefore the middle of the parallel piece would come in the same place.

To get the screw sufficiently below the water, the difference of draught of water may be about 2 feet, and consequently the depth from the water-line to the rabbat of the keel on the aftermost perpendicular = 19·087, and on the foremost perpendicular = 17·087.

Log b = 1·65531
Log 5·6 = 6·74819

Log $\dfrac{b}{5\cdot6}$ = 0·90712 and 8·075, or 8 feet the height of the portsill above the load water-line.

As the last example, suppose a river steam-boat is to be built of iron; that it shall have a steam-engine of 120 nominal horse power, and carry coals for 12 hours' full steaming; that it per medium will have on board 350 passengers and a complement of 12 men.

The displacement is calculated as follows:—

Steam-engines of 120 nominal horse power = 120 × 0·71 = .   85·2 tons
Coals for 12 hours = $\dfrac{120 \times 0\cdot15}{2}$ = ...   90   „

300 passengers with goods $= 350 \times 0\cdot1$ . . . . . 35·0 tons
12 men belonging to the ship, with clothes, &c. $= 12 \times 0\cdot11 = $ 1·32 ,,
Weight of hull, with anchors, chains, &c. $= 0\cdot836\ s = $. 109·12 ,,

Total displacement $= \overline{239\cdot64}$ tons
which, multiplied by 35, gives the displacement $=$ D $= 8387\cdot4$ cubic feet.

As this boat is intended for smooth water only, the length of the construction water-line may be 9·5 times the breadth, that is $l = 9\cdot5$ B, and the depth from the water-line to the keel may be one-fifth of the breadth; that is, $d = 0\cdot2$B $- m$ may be $= 0\cdot95$ and $n = 2$.

It is thought advisable to give the vessel a parallel piece of a length $= 3\cdot5$ B; the remainder of the length l will then be $= 6$B, and the displacement D $= \dfrac{n}{n+1}$ $\times 6$ B $\times \phi + 3\cdot5$ B $\times \phi$; and as ϕ is $= m$ B $d = 0\cdot2 \times 0\cdot95$ B $^3 = 0\cdot19$ B ^{3}, D is $= 0\cdot19$ B $^3\left(6 \times \dfrac{n}{n+1} + 3\cdot5\right)$ $= 0\cdot19$ B ^{3} $(4+3\cdot5) = 1\cdot425$ B ^{3}, and B $= \sqrt[3]{\dfrac{D}{1\cdot425}}$.

B is then found to be $= 18\cdot055, l = 171\cdot527$, and if the addition to the construction water-line is made $= 4\cdot473$ (2·473 forward and 2 aft) the length between the perpendiculars will be $= $ L $= 176, d = 3\cdot611, h = 3\cdot43, \phi = 61\cdot95$.

Length of parallel piece $= 63\cdot203$ and $l - 63\cdot203 = 108\cdot324$.

63·203 × 61·95 $=$ . . 3915
⅔ × 108·324 × 61·95 $=$ . . 4472·4
The displacement at above $= \overline{8387\cdot4}$ cubic feet.

From the after-end of the construction water-line to

D

the after-end of the parallel piece may
be about 2 B, or . ˙ . .   .   . 36 feet

Half of the parallel piece is $\dfrac{63\cdot203}{2} =$. 31·601 „

From the after-end of the construction
water-line to the middle of the parallel
piece. .   .   .   .   .   . 67·601 „

Half length of construction water-line
$= \dfrac{171\cdot527}{2} =$.   .   .   .   . 85·763 „

Middle of the parallel piece abaft the
middle of the construction water-line 18·162 „

which is $= \div \dfrac{(v\,n + 1)\,(n + 2)}{v\,n + 2}\,z$, v being $= \dfrac{l}{p}$ when P is
the length of the parallel piece. Therefore $+\,z =$
$\dfrac{18\cdot162\,(v\,n + 2)}{(v\,n + 1)\,(n + 2)}\,\dfrac{l}{p} = \dfrac{171\cdot527}{63\cdot203} = 2\cdot714 = v$, and $-\,z =$
5·247, that is, the centre of gravity is 5·247 feet abaft
the middle of the construction water-line, and to this
circumstance the necessary consideration must be taken
by determining the place of the engines with boilers
and coals.

A steamer of this description needs not have any
difference of draught of water forward and aft. The
height from the water-line to the deck may be $= \dfrac{B}{3\cdot6} =$
5 feet.

The elements of construction of the different ships
here calculated are put down in the following table;
and according to these elements the drawings, Plates
III., IV., V., VI., VII., and VIII. are made up.

	Ship of 500 Tons.	Clipper of 500 Tons	Schooner of 94 Tons.	Ship of 1000 Tons.	Frigate of 59 Guns.	Steamer of 100 H.P.
Displacement = D . . . . cubic feet	279167	31500	8111	59420	126855	8387·4
Length of construction water-line = l . . . foot	110·89	131·71	68·79	144·02	248·7	171·53
Length between perpendiculars = L . . . ,,	113·39	134·21	71·29	146·52	258·0	176 0
Breadth moulded in water-line = B . . . ,,	30·27	34·74	20·86	37·31	45·22	18·06
Depth, from the load water-line to the rabbet of the keel or $\phi = d$ . . . feet	13·86	13·17	8·6	18·0	18·09	3·61
Area of midship section = ϕ . . square feet	352·4	3432	124·4	577·64	654·3	61·95
Greatest abscissa of line of section := h . . foot	11·64	9·88	6·02	15·48	14·47	3·43
Exponent of line of sections = n . . .	2·5	2·3	2·5	2·5	2·3	3·
Exponent of forepart of water-line = n^1 . .	5·152	4·746	4·	6·	5·744	
$\dfrac{\phi}{B\,d} = m$	0·84	0·75	0·7	0·86	0·8	0·95
Centre of gravity before the middle of the water-line . feet	2·268	2·684	1·4258	2·93	·÷1·65	—·549
Do. before do. of construction water-line . . . ,,	2·518	2·934	1·676	3·18	0·	—·5·25
ϕ before do. of do. . . . . ,,	11·33	12·617	7·541	14·312		0·
Difference of draught of water ·k· . . ,,	1·25	1·25	1·25	1·25	2·	
Depth from the water-line to the rabbat of keel abaft ,,	14·61	13·91	9·35	18·75	10·09	3·61
Do. do. forward ,,	13 36	12·66	8·1	17·5	17·09 to portsill	3·61
Height from the water-line to the deck . . ,,	5·83	6·5	...	...	8·	5·
Length of parallel piece = r . . . ,,	...	...	...	...	67·83	69·2
Middle of parallel piece abaft middle of construction water-line . . . feet	...	...	...	...	0·	18·16
$\dfrac{L}{B} =$	3·746	3·864	3·45	3·93	5·705	9·745
$\dfrac{d}{B} =$	0·458	0·379	0·416	0·482	0·4	0 2
$\dfrac{D}{L \times B \times d} =$	0·587	0·513	0·5	0·604	0·601	0·531

CHAPTER V.

*To make up a Drawing to a Ship according to the
calculated elements of construction.*

From the calculated elements of construction in the
foregoing Chapter, we will select those for the common
merchant ship of 500 tons.

First construct a scale, which in general may be $\frac{1}{4}$ of
an inch to a foot, or, which is the same thing, $\frac{1}{48}$th of
the true size (that the plates should not be too large, a
smaller scale is here used). Then draw a line $a\,b = l$,
$= 113\cdot30$ feet (Pl. III). Put $a\,c = 1$ and $b\,d = 1\cdot5$
and $c\,d$ is $= l = 110\cdot89 =$ the length of the construction water-line. Draw perpendiculars to the line $a\,b$
from a and b, and put $a\,a' = b\,b'$ a little more than the
draught of water abaft, $a'\,a'' = b'\,b''$ a little more than
the half breadth. Draw $a'\,b'$ and $a''\,b''$. Then make
$a\,\phi = a''\,\phi'' = 67\cdot775$ feet, and ϕ is the place for the
midship section or dead-flat. Put $a'\,c' = a\,c$ and $b'\,d'$
$= b\,d$. Divide $c\,\phi$, $c'\,\phi'$, $d\,\phi$, and $d'\,\phi'$ each into 10
equal parts, and draw the lines 1, 2, 3, 4, 5, 6, 7, 8, 9.
Divide, likewise, $a'\,\phi'$, $a''\,\phi''$, $b'\,\phi'$, and $b''\,\phi''$, each into 10
equal parts, and draw the lines 1', 2', 3', 4', 5', 6', 7', 8',
9', through the corresponding points on the forepart,
and the line 7' through the seventh part on the after-

part. Put $a\,c = 14\cdot608$, $\phi\,m = 13\cdot861 = d$, and $b\,n =$ 13·358, and draw $o\,m\,n$, which is the upper edge of the rahbat of the keel, and which is a straight line if the calculations and the drawing are correctly executed.

To find the ordinates of the line of sections, the figure on Plate II. is to be used. From the mark 2·5 on the line A B as a centre, and with $h = 11\cdot643$ as a radius, describe part of a circle ef. From the other end of the line marked, 2·5, draw a straight line gh, just touching the circle, and the distances from the points 1', 2', 3' 4', &c., perpendicular to the line gh are the ordinates for the line of sections for the corresponding vertical sections of the ship. These ordinates multiplied by the main breadth $n = 30\cdot27$ give the areas of the sections. However, these areas may be found without any calculations whatever, by applying the area of the midship section $= \phi = 352\cdot43$, taken on a convenable scale A (Plate III.) as a radius, instead of h, and the same point as a centre, and the distances from the points 1', 2', 3', 4', &c., perpendicular to the line gh, figure, Plate II., measured on the same scale A, will give the area of the corresponding sections. If these distances are applied as ordinates on the corresponding sections 1—2—3, and (Plate III.) from the water-line downwards, and a curved line $c\,k\,d$ is drawn through all the points, the area of every section between c and d, may be found by measuring the distance from the curved line $c\,k\,d$ to the water-line ab on the same scale A, on which the area of the midship section was taken. Draw now the sections or frames that are to be used in the actual building of the ship, viz.: 3—9—12—15, &c. on the afterparts, and c—F—I—M, &c. on the forepart, and measure their areas as follows:—

Before φ			Abaft r		
φ .	. 852·4	square feet	3 .	. 351·2	square feet
c .	. 349·0	,,	6 .	. 347·0	,,
r .	. 338·3	,,	9 .	. 339·0	,,
I .	. 315·0	,,	12 .	. 326·0	,,
M .	. 277·0	·,	15 .	. 305·2	,,
p .	. 2·9 5	,,	18 .	. 277·2	,,
s .	. 142 2	—,,	21 .	. 242·5	,,
			24 .	. 200·0	,,
			27 .	. 147·1	,,

Now put $\phi'g' = \frac{1}{2}$B$=15\cdot195$ and apply this distance on the same figure, Plate II., and in the same manner as h was applied for finding the ordinates of the line of sections, but from a line kl drawn parallel with A c at the point $5\cdot152 = n' = $ the exponent of the water-line marked from the scale underneath the figure. Then the breadths of the load water-line for the sections 1', 2', 3', &c. on the fore-body, Plate III., and for the section 7' on the after-body, are found in the same manner as were the ordinates of the line of sections. The stern, and stern-post may now be drawn according to the determined rake. The height ϕn from the water-line to the deck is put up $= 5\cdot83$ feet, and the gunwale, rails, stern and head may be drawn. Now the parallelogram 1, 1, 7, 7 is drawn, into which 1, 1 is $=$ B, and 1, $7 = d$; 7, $q = q$, $7 = 1$, $r = r$, $1 = \frac{B}{2}$. Put then 7 i $= 2d (1—m) = 4\cdot4355$, and the trapezium $l r q i$ will be equal to half the area of the midship section. Divide

the line $q\,i$ into six equal parts, mark the thickness of
the keel, and draw a well continued fair curved line
from the upper edge of the rahbat of the keel through
the points 1 and 5 of the line qi to the point 1 of the
water-line 1, 1. This will be the midship section, and
has nearly the determined area, and may easily be
corrected to hold exactly the number of square feet
before calculated for the midship section. By this
method the midship section will obtain the form that,
with the given breadth, depth and area, will produce
the greatest stability. Now the top breadth lines must
be drawn in the longitudinal horizontal plan, and thereafter all the transverse sections or frames. In the
forepart is given, for each transverse section, the area,
the rahbat of the keel, the breadth in the water-line,
and the breadth at the gunwale; in the afterpart the
same, except the breadth in the water-line, which only
is nearly known for the section that comes nearest to
the section 7′. Not everybody will be able instantly to
draw the different sections to their determined areas
without many corrections, but after some practice it may
be done without any great trouble; and when the areas of
the sections are kept exactly equal to their predetermined areas, the designer is sure of having got his ship's
body to answer to the calculated elements of construction. This may he ascertained, and the height of the
centre of gravity of the displacement found, as well
as the place of the metacentre, by such calculations
as are made in the following Chapter, from measure-
ments on the drawing. Such a high degree of accuracy

as is sought in this example is nevertheless not necessary; for in a displacement of nearly 30,000 cubic feet, a difference of some hundred cubic feet is of no consequence.

If anybody should be of opinion that the stability of the ship is not known to a sufficient degree of accuracy by the place of the metacentre, they may use the method laid down by Atwood (See Phil. Trans. of Royal Society of London, 1796 and 1798). However, the method of determining the stability by the place of the metacentre may be considered sufficiently accurate in all cases where the sides of the ship, a little above as well as below the water-line, are parallel with each other, or nearly so.

CHAPTER VI.

*To calculate the Displacement, the Place of its Centre
of Gravity; and of the Metacentre.*

FOR these calculations we will use the drawing for
the same ship of 500 tons, Plate III.

To find the areas of the transverse sections.

The distance between the water-lines on the drawing is 2·31, and one-third of this distance is = 0·77.
feet.

Section Z.

No. of Water-lines	Ord.		
1	6·85	1	6·65
2	5·00	4	20·00
3	3·10	2	6·2
4	1·42	4	5·68
5	0·35	1	0·35
			38·88
			0·77
		⅓ Area = 29·94	

Section 8.

No. of Water-lines	Ord.		
1	10·9	1	10·7
2	9·4	4	37·5
3	7·38	2	14·76
4	4·92	4	19·68
5	2·5	2	5·0
6	0·80	4	3·44
7	0·25	1	0·25
			91·63
			0·77
		⅓ Area = 70·63	

Section M.

Ord.		
14·68	1	14·56
14·27	4	57·08
13·65	3	27·10
11·88	4	47·52
8·77	2	17·64
4·03	4	16·12
0·25	1	0·25
		180·19
		0·77
⅓ Area = 138·75		

Section F.

Ord.		
15·05	1	15·05
15·02	4	60·08
15·00	2	30·00
14·60	4	58·40
12·85	2	25·70
7·52	4	30·08
0·30	1	0·30
		219·61
		0·77
⅓ Area = 169·1		

Section 7.

Ord.		
15·13	1	15·13
15·13	4	60·52
15·13	2	30·26
14·90	4	59·60
13·61	2	27·22
8·91	4	35·64
0·45	1	0·45
		228·82
		0·77
⅓ Area = 176·19		

Section 6.

Ord.		
15·13	1	15·13
15·12	4	60·48
15·00	2	30·00
14·75	4	59·00
13·25	2	26·50
8·65	4	34·60
0·50	1	0·50
		226·21
		0·77
⅓ Area = 174·18		

Section 12.

No. of Water-lines	Ord.		
1	14·80	1	14·80
2	14·75	4	59·00
3	14·55	2	29·10
4	14·00	4	56·00
5	12·30	2	24·60
6	8·75	4	27·00
7	0·50	1	0·50
			211·00
			0·77
		⅓ Area = 162·47	

Section 13.

Ord.		
14·20	1	14·20
14·00	4	56·00
13·60	2	27·20
12·35	4	49·40
9·20	2	18·40
3·45	4	13·80
0·50	1	0·50
		179·50
		0·77
⅓ Area = 138·22		

Section 24.

Ord.		
12·90	1	12·90
12·38	4	49·52
10·95	2	21·90
7·96	4	31·84
3·95	2	7·90
1·25	4	5·00
0·60	1	0·60
		129·56
		0·77
⅓ Area = 99·76		

Section 30.

Ord.		
8·95	1	8·95
6·30	4	25·20
3·55	2	7·10
1·92	4	7·68
1·00	2	2·00
0·50	4	2·00
0·50	1	0·60
		53·43
		0·77
⅓ Area = 41·14		

Section 22.

Ord.		
4·80	1	4·80
2·50	4	10·00
1·45	2	2·90
0·93	4	3·72
0·57	2	1·14
0·60	4	2·00
0·50	1	0·50
		25·06
		0·77
⅓ Area = 19·3		

To find the cubical content and centre of gravity of the part from section 30 to the aftermost perpendicular :
From section 30 to the aftermost perpendicular is 7·77 feet, therefore the distance between the section 3·885, and one·third of this distance = 1·295 feet.

Section.	½ Area.				
30	41·14	1	41·14	0	
32	19·30	4	77·20	1	77·2
Strnpst.	0·0	1	0·00	2	

118·34   77·2
1·295   3·835

½ Cubical cntnt. 153·25   118 34)299·92

2·531 Centre of gravity from section 30.
7·77 Sect. 30, from aftermost perpendiclr,

5·246 Centre of gravity from do.

To find the cubical content and centre of gravity of the part from section 30 to section S : Distance between the sections = 12, and one-third of this distance = 4 feet.

Section.	½ Area.				
30	41·14	1	41·14	0	
24	99·76	4	399·04	1	399·04
18	138·22	2	276·44	2	552 88
12	162·47	4	649·88	3	1949·64
6	174·19	2	348 38	4	1393·52
✦	176·19	4	704 76	5	3523·80
F	169·10	2	338·20	6	2029·20
M	138·75	4	555·00	7	3885·00
S	70·56	1	70·56	8	564·48

3353·40)14297·56
4

4·2258
½ Cub. cntnt. 13533·6   12

50·7096 Centre of gravity from section 30.
7 77 Section 30 from aftermost perpendclr.

58 4796 Centre of gravity from   do.

To find the cubical content and centre of gravity of
the part from section S to the foremost perpendicular :
From section S to the foremost perpendicular is 9·62 ;
distance between the sections 4·81, and one-third of
this distance = 1·603 feet.

Section. ¼ Area.
```
 S   '70·50 1  70·56 0
 Z    29·94 4 119·76 1          119·76
Stem   0·00 1   0·00 2

          190·32              119·76
            1·603               4·81

¼ Cub. cntnt.  305·08   190·32)576·05
```

3·026 Centre of gravity from Sect. S.
103·77 Sect. S. from aftmst. perpndclr.

106·796 Centre of gravity from   do.

To find the whole displacement D.

```
¼ Cubical content from section S0 to the stern-post      153·25
      „        „        „     30 to section S  .   13533·60'
      „        „        „      S to the stem  .  .    305·08

             ¼ Displacement  .  .  .  .  .  .   13991·93
                                                      2

             The whole Displacement = D =   .  .   27983·86 cub. ft.
```

To find the centre of gravity of the whole displacement.

This centre is from the aftermost perpendicular :

$$\frac{305·08 \times 106·796 + 13533·6 \times 58·4796 + 153·25 \times 5·246}{13991·93} = . \quad 58·95$$

Half of the length between the perpendiculars . . 56·695

Centre of gravity before the middle of the load water-line   2·255 feet

The displacement should have been 27916·7 cubic

feet, and the centre of gravity of the displacement 2·268 feet before the middle of the load water-line; consequently the displacement has been 67·16 cubic feet too large, and the centre of gravity has come 0·013 feet = 0·156 inches too far abaft; but this small incorrectness is of no consequence in a machine of this magnitude, and might easily have been obviated if the drawing had been made to a larger scale.

To find the area of the water-lines.

The distance between the sections is 12 feet, and one-third of this distance is = 4 feet.

Sections.	1st Water-line Ord.			2nd Water-line Ord			3rd Water-line Ord.			4th Water-line Ord.		
S	10·90	1	10 90	9·40	1	9 40	7·38	1	7·38	4·92	1	4·92
M	14·58	4	58·32	14·27	4	57·08	13·55	4	54·20	11·88	4	47·52
F	15·05	2	30·10	15·02	2	30·04	15·00	2	30 00	14·60	2	29·20
✦	15·13	4	60·52	15·13	4	60·52	15·13	4	60·52	14·90	4	59·60
6	15·13	2	30·26	15·12	2	30·24	15·00	2	30·00	14·75	2	29·50
12	14·80	4	59·20	14·75	4	59·00	14·55	4	58 20	14·00	4	56·00
18	14·20	2	28·40	14·00	2	28·00	13·60	2	27·20	12·35	2	24·70
24	12·90	4	51·60	12·38	4	49·52	10·95	4	43·80	7·96	4	31·84
30	8·95	1	8·95	6·30	1	4·30	3·55	1	3·55	1·92	1	1·92
			338·25			330·10			314·85			285·20
			4			4			4			4
			1353·00			1320·4			1259·4			1140 8
Before Sect. S			60·59			47·13			31·71			17·95
Abaft „ 30			30 84			21·49			12·43			7·53
⅓ Area=1450·43				⅓ Ar.=1389·02			⅓ Ar.=1303·64			⅓ Ar.=1166·28		

Sections.	5th Water-line.			6th Water-line			7th Water line.		
	Ord.			Ord.			Ord.		
S	2·50	1	2·50	0·66	1	0·66	0·25	1	0·25
M	8·77	4	35·08	4·03	4	16·12	0·25	4	1 00
F	12·85	2	25 70	7·52	2	15·04	0·30	2	0·60
✦	13·61	4	54·44	8·91	4	35 64	0·45	4	1·80
6	13·25	2	26·50	8·65	2	17·30	0·50	2	1·00
12	12·30	4	49·20	6·75	4	27·00	0 50	4	2·00
18	9·20	2	18·40	3·45	2	6·90	0·50	2	1·00
24	3 95	4	15·80	1·25	4	5·00	0·50	4	2·00
30	1·00	1	1·00	0·50	1	0·50	0·50	1	0·50
			228·62			124·06			10·15
			4			4			4
			914 48			496·24			40·60
Before Sect. 8			6 97			2·44			0·00
Abaft ,, 30			4·36			1·17			1·1
	½ Area= 925·81			½ Ar.= 499·85			½ Ar.= 41·7		

To find the displacement and the place of the centre of gravity below the load water-line.

Water-lines.	½ Area of Water-lines.				
1	1450·43	1	1450·43	0	
2	1389·02	4	5556·08	1	5565·08
3	1303·54	2	2607·08	2	5214·16
4	1166·28	4	4665·12	3	13995·36
5	925·81	2	1851·62	4	7406·48
6	499·85	4	1999·40	5	9997·00
7	41·70	1	41·70	6	250·20
			18171·43		)42419·28
			·77		
					2·334
½ Displacement =			18992·00		2·31
			2		
					5·391 Centre of gravity of dis-
Whole dsplcmt =D= 27984·0					placement below the load water-line.

To find the metacentre.

Sects.	Ordinates of first Water-line.	Cubes of Ordinates.		
S	10·9	1295·03	1	1295·03
M	14·8	3112·14	4	12448·56
F	15·1	3442·95	2	6885·90
♣	15·1	3442·95	4	13771·80
6	15·1	3442·95	2	6885·90
12	14·8	3241·79	4	12967·16
18	14·2	2863·29	2	5726·58
24	12·9	2146·69	4	8586·76
30	9·0	729·00	1	729·00

$$\begin{array}{r} 69296·69 \\ 4 \\ \hline 277186·76 \end{array}$$

The triangle before sect. S   3114·55
   ,,   abaft   ,,   30   1416 08

$$\begin{array}{r} 281717·39 \\ 2 \\ \hline 3)563434·78 \\ \hline 27994)197611·593 \end{array}$$

6·711 Metacentre above the cen. of gr. of D.
5·239 Centre of gr. below the load water-ln.

1·320 Metacentre above the load water-line

CHAPTER VII.

On the Proportion of Sails.

A SHIP is not a complete machine before she has got her masts, spars, rigging, and sails. To get this machine as perfect and effective as possible, it is necessary that the masts, rigging, and sails shall have certain proportions to the size and stability of the hull —and consequently that they shall be made according to just principles.

If the place of the centre of gravity of the ship with everything in it is known with regard to the height above the keel, and this centre, as for instance in men-of-war, may be considered to be nearly in the same place during most of the time that the ship sails, the stability of the ship is almost unaltered for such a time, and the moment of stability for a given inclination may be calculated. As the pressure of the wind on a square foot is known for the different denominations of the strength of the wind, it may be determined that a ship at a certain state of wind, for instance, strong breeze, with certain sails set, and close hauled, shall incline a suitable number of degrees, for instance 7 or 8, and the moment of sails then be made equal to the moment of the stability. This

method ought to be applied to men-of-war without
steam power; but now that all men-of-war are provided
with this propelling power, the sails are generally
made smaller than the ship could otherwise carry,
so that the resistance of the rigging and spars,
when steaming against the wind, should not diminish
the ship's velocity too much. As for merchant ships,
they are obliged, at different times, to sail with very
different draughts of water, according to the weight of
the cargo or ballast, by which circumstance the stability becomes extremely variable. It is therefore
impossible to give to a merchant ship sails of the most
suitable size for all circumstances, and the best plan
to proceed on will therefore be for the sails of merchant ships to use such proportions as by practice are
found the most suitable.

It ought to be remarked, that as the breadth has
most influence on the stability, the masts, topmasts,
and topgallantmasts reasonably ought to have a certain
proportion to the breadth, by which the height of
every sail, as well as the height above the water-line
of their common centre of gravity, will be in proportion to the ship's breadth; but the length of the yards
is generally made in a certain proportion to the length
of the ship, by which the moment of sails will be in
proportion to the length multiplied by the square of
the breadth. Small vessels will then get a greater
moment of sails in proportion to their stability than
large ones; but, when a sudden squall comes on, every
necessary manoeuvre for the safety of the ship may also

he done in a shorter time in the small vessel than in the large one.

It may also be recommended to observe, that all the masts and spars shall have a certain proportion to each other, as well as to the dimensions of the ship, in order that the whole may get a handsome appearance; and it is a general notion that a ship is handsomely rigged when : —

1. The main-topmast-stay and the forestay make a straight line together, and are parallel to the mainstay, or nearly so. The forestay may come between $\frac{1}{3}$ and $\frac{2}{5}$ths of the length of the bowsprit outside the stem from the outer end of it.

2. That the three topsails are homologous, or at least that their sides are parallel.

3. That when the ship is seen from one of the ends, all the shrouds of the different masts should seem to be parallel to each other, or nearly so, and likewise the top-mast shrouds. This depends much on the breadths of the channels, that ought to be modified thereafter.

Besides this, the place of the masts with regard to the length of the ship is of consequence, and must be determined in such a manner that the ship may get a suitable ardency, that it may go about with facility, and that not too much of the power of the rudder may be necessary to keep the ship to her course when close-hauled. The science of naval architecture is not so far advanced that it is possible to calculate where the centre of effort of sails shall be to give the best steering, at least not with the desirable degree of accuracy ; and

therefore the ship's designer, in placing the masts, is left
entirely to what by experience is proved to answer best.
By this is found that the common centre of effort of
sails, that is their common centre of gravity, with pro-
priety may be from $\frac{1}{10}$ to $\frac{1}{20}$ of the ship's length in the
load water-line before the vertical passing through the
centre of gravity of the displacement; but in vessels
with a considerable difference of draught of water, such
as cutters and schooners generally are, it is commonly
near or abaft this said vertical.   In ships of great
length in proportion to their breadths, it seems to be
of less consequence if the centre of effort of sails is a
little farther forward or aft.

For men-of-war with steam·power, the same rules
for placing the masts, and the same proportions for
masts and spars, are applicable, only that, as the length
of these ships generally is great in proportion to the
breadth, the yards would be too long if made in the
same proportion to the length as for sailing merchant
ships.   Instead, therefore, of using the real length in
proportioning the yards, gaffs, &c., four times the
breadth may be considered as the ship's length.

As the place of the masts, according to the foregoing,
must be different for different shapes of ships' bottoms,
the following rules may be a good guidance, showing
how the masts in practice have been placed in regard
to the length of the ship.

If the length between the perpendiculars is $= 1$, the
place of the masts is found to have been for

FRIGATE-RIGGED SHIPS.

Foremast from the foremost perpendicular			0·135 L to 0 108 L.
Mainmast	do.	do.	0·57 L to 0·542 L.
Mizenmast	do.	do.	0·828 L to 0·355 L.

BARQUE SHIPS.

Foremast from the foremost perpendicular			0·14 L to 0·185 L.
Mainmast	do.	do.	0·58 L to 0·6 L.
Mizenmast	do.	do.	0·672 L to 0·816 L.

BRIGS.

Foremast from the foremost perpendicular			0·164 L to 0·196 L.
Mainmast	do.	do.	0·621 L to 0·6 L.

BRIG FORWARD AND SCHOONER AFT.

Foremast from the foremost perpendicular			0·154 L to 0·168 L.
Mainmast	do.	do.	0·61 L.

SCHOONERS.

Foremast from the foremost perpendicular			0·1902 L to 0·25 L.
Mainmast	do.	do.	0·64 L to 0 605 L.

KETCHES.

Mainmast from the foremost perpendicular			0·3333 L.
Mizenmast	do.	do.	0·82 L.

CUTTERS.

The mast from the foremost perpendicular	0·334 L to 0·371 L.

Rules for the Proportions of Masts and Spars.

Length in the load water-line or between the perpendiculars = L.

Greatest breadth moulded in the load water-line = B.

	Maximum	Minimum.
1. FOR FRIGATE-RIGGED SHIPS. *Mainmast* above the load water-line, head included . . . . . . The length below the water-line according to the depth of the ship.	2 077 B = A	1·92 B = A

	Maximum.	Minimum.
Mainmast-head	0·164 A	0·173 A
Main-topmast from tho lower side of the fid-hole, head included . . .	1·38 B = C	1·265 B = C
Main-topmast-head	0·117 C	0·117 C
Main-topgallantmast from the lower side of the fid-hole to the stops . . .	0·731 B	0·728 B
The length of the pole equal to the length from the topmast-cap to the stops of the topgallantmast, ⅛ to ¼ of this length to be above the royals.		
The head of the foremast below the horizontal line through the head of the mainmast, ⅔ of the length of tho mainmast-head, and the length of the foremast-head equal to the length of the mainmast-head multiplied by . .	0·88	1·0
The length of all the spars of the foremast, yards included, equal to the length of the corresponding spars of the mainmast multiplied by . .	0·88	1·0
The head of the mizenmast in a straight line drawn through the main and foretops.		
Nota — In barques the tresseltrees of the mizenmast shall be in this line.		
Mizenmast-head equal to the mainmast-head multiplied by	0·75	0·66
All the spars of the mizenmast, yards included, equal to the corresponding spars of the mainmast multiplied by .	0·75	0·66
Bowsprit outside the stem . . .	1·06 B	1·0 B
Jibboom outside the bowsprit cap . .	0·66 B	0·5 B
Spanker-boom whole length . . .	0·355 L = D	= 0·355 L = D
Gaff to the clamps	0·6 D	0·64 D
The middle of the lower yards below the upper side of the lower tresseltrees, ⅓ of the length of the mast head.		
The middle of the topsail-yards below the topmast-tresseltrees, ₁/₁₀ of the length of the topmast from the lower tresseltrees to the topmast tresseltrees.		
The middle of the topgallant-yards below the stops, ₁/₁₀ of the length of the topgallantmast from the stops to the topmast-tresseltrees.		
Breadth of main-topsail at the foot or the length of the main-yard, yard-arms excluded	0·4784 L = B	0·4615 L = B

	Maximum.	Minimum.
Breadth of main-topgallantsail at the head	0.556 x $=$ F	0.51 B $=$ F
Breadth of main-royals at the head .	0.74 F	0.68 F
Yard-arms of lower, topgallant, and royal yards, $\frac{1}{18}$ of the length of the yards, yard-arms excluded, each.		
Yard-arms of main topsail-yards, each .	$\frac{E-F}{6}$	$\frac{E-F}{6}$
Jib along the stay equal to the length of the stay multiplied by	$0.8 = P$	$0.75 = P$
The after leech	0.71 P	0.703 P
The foot	0.28 L	0.251 L
Spanker-gaff parallel to the mizen-stay.		
Rake of foremast in 12 ft. from 3 to 6 in,		
„ mainmast „ 6 to 12 „		
„ mizenmast „ 12 to 24 „		
Stive of bowsprit „ 2¼ to 5 ft.		
The length of the studdingsail-booms is the half of the length of the yards to which they belong.		

2. FOR BRIGS.

Mainmast above the water-line, head included	2.0585 B $=$ A	
Mainmast-head	0.1686 A	
Main-topmast, head included . . .	1.36 B $=$ C	
„ head	0.131 C	
Main-topgallantmasts to the stops . .	0.735 B	
Pole, as on frigates.		
The head of the foremast below a horizontal line through the head of the mainmast, ¼ the length of the main-mast-head.		
Length of main-yard, yard-arms excluded	0.51 L $=$ D	
Length of main-topgallant-yard, yard-arms excluded	0.5 D $=$ E	
Length of main-royal-yard, yard-arms excluded	0.76 E	
All spars of the foremast equal to the corresponding spars of the mainmast.		
The yards are hung as on frigates.		
Yard-arms of topsail-yards each . .	$\frac{D-E}{6}$	
All the other yard-arms as on frigates.		
Bowsprit and jibboom as on frigates.		
Main-boom, whole length	0.53 L $=$ F	
Gaff to the clamps	0.7 F	

	Maximum.	Minimum.
Jib along the stay equal the length of the stay multiplied by	0·8 = P	
After leech	0·72 P	
Foot	0·3 L	
Rake of foremast in 12 feet from 4 to 8 in.		
„ mainmast „ „ 8 „ 16 „		
Stive of bowsprit „ „ 9 „ 4 ft.		

3. BRIG FORWARD AND SCHOONER AFT.

	Maximum.	Minimum.
Foremast above the water-line, head included	1·9764 B = A	1·89 B = A
„ head	0·164 A	0·164 A
Fore-topmast, head included . . .	1·4178 B = C	1·35 B = C
„ head	0·1233 C	0·1233 C
Fore-topgallantmast to the stops . .	0·7 B	0·6676 B
Pole, as on frigates.		
Mainmast above the water-line, head included	2·9423 B = D	2·6 B = D
„ head	0·118 D	0·118 D
Main-topmast to the stops . . .	1·765 B	1·565 B
Pole above the stops equal to the pole above the royal on the fore-topmast.		
Bowsprit outside the stem . . .	1·0 B	0·75 B
Jibboom outside the bowsprit cap . .	0·75 B	0·66 B
Breadth of fore-topsail at the foot . .	0·52 L = E	0·5 L = E
Breadth of fore-topgallantsail at the head	0·5 B = F	0·425 E = F
Breadth of the royal at the head . .	0·745 F	0·745 F
The yards are hung as on brigs, and the length of all the yard-arms as on brigs		
Main boom, whole length . . .	0·58 L = D	0·5 L = D
Main-gaff to the clamps	0·7 D	0·603 D
Fore-gaff	0·3 L	0·28 L
„ below tresseltrees . . .	0·156 B	0·156 B
Main-gaff do. . . .	0·262 B	0·262 B
Rake of foremast in 12 ft. from 15 to 20 in.		
„ mainmast „ „ 24 „ 30 „		
Stive of bowsprit „ „ 80 „ 42 „		

4. FOR SCHOONERS.

	Maximum.	Minimum.
Foremast above the water-line, head included	2·622 B = A	2·5 B = A
„ head	0·143 A	0·143 A
Fore-topmast to the stops . . .	1·3 B	1·04 B
Pole above the stops, equal to the length of the topmast from the lower cap to the stops : ⅓ to ½ of this length should be above the topgallant sail.		

	Maximum.	Minimum.
Mainmast above the water-line, head included	$2.9423 B = D$	$2.6 B = D$
„ head	$0.118 D$	$0.118 D$
Main-topmast to the stops	$1.765 B$	$1.6 B$
Pole above the stops, equal to the pole above the topgallantsail on the fore-topmast.		
Bowsprit, outside the stem	$0.75 B$	$0.53 D$
Jibboom, outside the bowsprit-cap	$1.02 B$	$0.8 B$
Breadth of the topsail at the foot	$0.578 L = C$	$0.515 L = C$
Breadth of the topgallantsail at the head	$0.5 C$	$0.47 C$
The yards are hung, and the lengths of the yard-arms are as on the foregoing.		
Main-boom whole length,	$0.58 L = E$	$0.53 L = E$
Main gaff to the clamps	$0.7 E$	$0.555 B$
Fore-gaff, whole length	$0.322 L$	$0.322 L$
„ below the trestletrees	$0.28 B$	$0.28 B$
Main-gaff do. do.	$0.2 B$	$0.262 B$
Rake of foremast in 12 feet, 12 to 20 in.		
„ mainmast „ 13 „ 30 „		
Stivo of bowsprit „ 24 „ 80 „		

5. FOR KETCHES.	1st Example.	2nd Example.
Mainmast, whole length	$3 B = C$	$3.2 B = C$
„ head	$0.166 C = T$	$0.21 C = T$
Main-topmast hounded	$1.264 B = D$	$1.5 B = D$
„ headed	$0.704 D$	$0.46 D$
Of this head or pole the royal is	$\frac{1}{4}$	$\frac{1}{4}$
Mizenmast-head	in haight with the mainmast-head.	in haight with $\frac{3}{4}$ of the main-mast-head.
Mizenmast-head, long	$\frac{3}{3} T$	$\frac{1}{4} T$
Mizen-topmast to the stops in height, with main-topmast hounded.		
Pole	equal to the main-topmast pole without the royal.	equal to $\frac{3}{4}$ of the mizen-top-mast pole without the royal.
Bowsprit outside the stem	$0.76 B$	$1.26 B$
Jibboom outside the bowsprit-cap	$0.76 B$	$0.76 B$
Fore-gaff	$0.341 L$	$0.4 L$
Driver-gaff to the stops	$0.3 L$	$0.24 L$
Driver-boom	$0.47 L$	$0.451 L$
Breadth of topsail at the foot	$0.568 L = E$	$0.618 L = B$
„ topgallantsail at the head	$0.477 B = F$	$0.55 B = F$
„ the royal at the head		$0.7 F$

	1st Example.	2nd Exampl's.
The fore-yard is hung below the tressel-trees	0·447 T	0·565 T
The topsail-yard below the stops . .	0·1	
„ below the upper ridge of the lower cap	. .	0·172 T
The gaffs below the tresseltrees . .	0 4 T	0·833 r
Yard-arms of the fore yard, top-gallant-yard, and royal-yard ¹⁄₆ of the breadth of the sails each.		
Yard arms of topsail yard	$\frac{E - F}{\sigma 7}$	$\frac{E - F}{6}$
Rake of mainmast in 12 feet . 5 inches.		
„ mizenmast „ . 12 „		
Stive of bowsprit „ . 32 „		

6. FOR CUTTERS, YACHTS, AND REVENUE CRUISERS.

	With Square Sail A	Without Square Sails	Yachts		Revenue Cruiser	
			Example 1.	Example 2.	Example 1.	Example 2.
Mast, whole length	3·6 B=C	3·39 B=C	4·01 B=C	3·56 B=C	3·16 B=C	3·2 B=C
„ head	0·21 C=D	0·167 C=D	0·167 C=D	0·186 C=D	0·167 C=D	0·187 C=D
Topmast bounded	1·5 B=A	2·0 B=A	2·3 B=A	1·85 B=A	1·79 B=A	1·72 B=A
„ headed	0·3 A	0·166 A	0·26 A	0·17 A	0·3 A	0·28 A
Bowsprit outside the stem	1·5 B	1·5 B	2·1 B	1·82 B	1·66 B	1·71 B
Main-boom, the length on deck multiplied by	0·863	0·92	0·963	0·955	0·92	0·87
Main-gaff. main boom, multiplied by	0·537	0·64	0·627	0·847	0·68	0·64
Breadth of topsail at the foot	0·7 L=E	0·752 L				
Squaresail-yard without yard-arms	0·55 E=F					
Breadth of topgallantsail at the head	0·7 F					
„ royal	0·5 D	0·55 D				
Squaresail-yard below trestletrees						
Topsail-yard below the cap ⁷⁄₁₀ of the mast-head	0·35 D	0·35 D				
Gaff below trestletrees						
Yard-arms as on ketches.						
Jib along the stay, the length of the stay multiplied by	0·975	0·875				
Fore-staysail along the stay, the length of stay multiplied by	0·9	0·9				
Rake of mast in 12 feet	10 inches	12 inches	12 inches	15 inches	14¾ inches	13 inches,
Size of bowsprit	10 „	10 „	7¾ „	10¼ „	18 „	16 „
The main-boom above the deck according to the size of the vessel, from 3 to 5 feet.						

According to these proportions the sails may he drawn, and thereafter the corrections made that may be found necessary to get the centre of effort of sails in the intended place, and that the whole may get a handsome appearance. In this manner the draughts of the sails are made in the Plates IX., X., XI., XII., and XIII.; for the ships on the Plates III., IV., V., VI., and VII.; and the place of the centres of effort of sails for the merchant ship Plate III. is found by the following calculations. In the same way the centre of effort of sails for the other ships may be found.

Sails.	Area, square feet.	Area of gravity above the water-line.	Moment.	Area of gravity from aftermost perpendicular.	Moment.
Driver	1022·87	27·10	27719·77	0·2	−204·57
Mizen topsail	760·80	54·00	41083·20	16·0	12172·80
Mizen-topgallantsail	350·04	75·25	26340·51	16·0	5600·64
Main-course	1781·60	26·80	47746·88	50·7	90327·12
Main-topsail	1480·60	63·00	93277·80	50·7	75066·42
Main-topgallantsail	686·86	92·70	63625·57	50·7	34798·45
Fore-course	1478·46	26·60	39327·03	100·89	149161·82
Fore-topsail	1194·00	58·90	70326·60	100·89	120462·66
Fore-topgallantsail	592·80	85·55	50714·04	100·89	59807·59
Fore-topmaststaysail	357·91	35·60	12741·59	126·99	45450·99
Jib	533·60	41·40	22091·04	136·39	72777·70

Area of sails . . 10239·4   )494994·03   665626·19

Centre of effort of sails above the water-line . . . . .)   48·34   204·57

10239·04)665421·62

Centre of effort of sails from the aftermost perpendicular . 64·98
Half length of the water-line . . . . . . 56·695

Centre of effort of sails before the middle of the water-line   8·285
Centre of gravity of displacement   ,,   ,,   . 2·555
Centre of effort of sails before the centre of gravity   .   . 5·73

which is $= \frac{L}{10\cdot 8}$, L being the length between the perpendiculars.

For the clipper ship, Plate IV., the area of sails is found 15710·4 sq. ft. The moment of sails $=$ 908886, and the centre of effort of sails before the centre of gravity of the displacement 7·54 feet, which is $= \frac{L}{17\cdot 8}$.

If the height of the metacentre above the centre of gravity of the displacement is $=$ H, and the displacement $=$ D, the product H D may be considered as a measure for the ship's stability. By experience it is found that the moment of sails in common merchant ships may be from 2·5 H D to 2·7 H D; and in clipper ships and men-of-war without steam power from 2·9 H D to 3·1 H D. In the merchant ship, Plate III., the moment of sails is 494994·03, which is nearly 2·6 H D; and for the clipper ship the moment of sails is 908886, which is about $=$ 3 H D.

CHAPTER VIII.

On the Scale of Capacity.

FOR every one that commands a ship it is of consequence to know how deep she will swim with a
certain predetermined cargo, as well as how much
deeper she will go when she has got some weight on
board and is to take in a certain quantity more; likewise what quantity of ballast it is necessary to take in
to get the ship down to a proper draught of water
when she has uone, or only a small cargo not sufficient
to give her the proper stability.

A sure and simple means of finding this is therefore
desirable, and this means gives the scale of capacity,
wherefore every ship by its constructor ought to be
provided with such a scale of capacity with a description of its use.

To show how a scale of capacity is drawn, the
necessary calculations are put down here for drawing a
scale of capacity for the ship, Plate III.

According to the calculations for the same ship,
Chapter VI.,—

The area of the 1st water-line is = ... 2900·86
"   2nd   do.   = · . . . . . 2778·04
                                        ⁴)5678·90
                                          2839·45
Distance between the water-lines . . . . . 2·31

Cubical content between 1st and 2nd water-line . . 6559 13
"   of planking between do. . . . 153·8
"   of stem and stern post between do. . 6·1
Whole cubical content between 1st and 2nd water-line . 6719·03

Area of the 1st water-line = 2900·86   1 . . . . 2900·86
"   2nd   do.   2778·04   4 . . . . 11112·16
"   3rd   do.   2607 08   1 . . . . 2607·08
                                            16620·10
                                             0·77
Cubical content between 1st and 3rd water-line . . . 12597·48
"   of planking between   do. . . . 383·8
"   of stem and stern-post between do. . . 12·1
Whole cubical content between 1st and 3rd water-line . 12943·38

Area of the 3rd water line . . . . . . 2607·08
"   4th   do. 2332 56
                                        ⁴)4939·64
                                          2469·82
Distance between the water-lines . . . . . 2·31
Cubical content between 3rd and 4th water-line . . . 5705·29
"   of planking between   do. . . . 176·4
"   of stem and stern-post between do. . . 6·0
Whole cubical content between 3rd and 4th water-line . 5887·69
"   "   1st and 3rd do. . . 12943·38
Whole cubical content between 1st and 4th water-line . 18831·07

In this way the cubical content between the first
and each of the other water-lines are as inserted
here :—

Cubical content between 1st and 2nd water-line . $6719.03 = a.$
    ,,   1st and 3rd do. . . . $12943.38 = b.$
    ,,   1st and 4th do. . . $18831.07 = c.$
    ,,   1st and 5th do. . . . $23921.46 = d.$
    ,,   1st and 6th do. . . $27496.04 = e.$
    ,,   1st and 7th do. . . . $29085.10 = f.$
Whole displacement, keel included . . . $29203.9 = y.$ ●

Now to construct a scale of capacity, draw first a scale A, Plate XIV., on which the whole displacement may be taken, and likewise a scale B, on which may be taken the draught of water of the ship. Draw a line $a\,b = 29203.9$ taken on the scale A. Draw the lines $a\,t$ and $b\,u$ perpendicular to $a\,b$. Make $a\,t = b\,u =$ the draught of water of the ship at the midship section, taken in feet and inches on the scale B. Now $a\,c$ is made $= c\,e = e\,g = g\,i = i\,l = l\,n = 2.31$, which is the distance between the water-lines taken on the scale B, and the lines $c\,d$, $c\,f$, $g\,h$, $i\,k$, $l\,m$, and $n\,o$ are drawn parallel to $a\,b$. Put now $c\,2 = a$ in the foregoing table, $e\,3 = b$, $g\,4 = c$, $i\,5 = d$, $l\,6 = e$, $n\,7 = f$, altogether taken on the scale A. Through the points a 2, 3, 4, 5, 6, 7 and u, draw a fair curved line, and the scale of capacity is completed.

A ton is equal to the weight of 35 cubic feet of sea-water, and therefore a scale of tons agreeable to the scale of capacity may be drawn by means of the scale A, and this is done in the scale D.

To explain the use of this scale of capacity we will suppose that the ship to which it belongs, with some weights on board, draws 11 ft. aft, and 9 ft. 8 in. forward. The mean draught of water is $= \dfrac{11\,\text{ft.} + 9\,\text{ft.}\,8\,\text{in.}}{2}$

$= 10$ ft. 4 in. It is desired to be known what will be the mean draught of water when an additional weight of 100 tons is taken on board. This is found as follows:—From u to a'' on the scale of capacity put 10 ft. 4 inches taken on the scale B, through a'' draw a line $a''\,b''$ parallel to $a\,b$, and through b'' a line $b''\,c''$ parallel to $b\,u$. From c'' put 100 tons taken on the scale D to d'', draw $d''\,e''$ parallel to $b\,u$ and $e''\,f''$ parallel to $a\,b$. The distance $f''\,u$ measured on the scale B, gives the ship's mean draught of water after the 100 tons are got on board, equal to 11 ft. 8½ in.

CHAPTER IX.

§ 1. On the dimensions of materials used in Ship-building.

THE strain that goes on the different parts of a ship it is impossible to calculate, and therefore it is by a long practice alone that the materials used in ship-building have got the dimensions now considered proper. It is to be observed that a ship intended to carry heavy goods, such as iron, ore, salt, &c., liable to make the ship uneasy at sea, ought to be built stronger than a ship intended only, for instance, for the timber trade. Different qualities of materials also require different dimensions to give the same strength to the ship, but still the strength does not depend alone on the dimensions given to the timbers and other materials. Their proper connections with suitable fastenings affect the strength of the ship almost in a still greater degree, and therefore it requires all the skill of an able and experienced practical ship-builder to give all the materials proper dimensions in proportion to each other, and to the dimensions of the ship.

The dimensions of the principal materials used in the building of men-of-war, as put down in the following table, are mean dimensions of what has commonly been used in different countries, and may therefore be considered as suitable in all common cases where the ship is built of oak, or some other timber of nearly the same strength; but if fir is applied to ship-building, the scantlings ought to be $\frac{1}{6}$ to $\frac{1}{5}$ greater. The same scantlings are applicable to merchant ships: it is only to be observed that the main-deck beams of a frigate will be suitable as lower-deck beams of a merchant ship of the same breadth, and the upper-deck beams of a frigate to the upper-deck beams of a merchant ship.

As a general rule it may also be observed that if the ship is very long in proportion to her breadth, the dimensions of some of the materials ought to be greater than put down in the table, and this about in the following proportions:—If the length in the load water-line is from 5 to 6½ times the main breadth, divide the length by 5, and take the quotient as the breadth for determining the scantlings of the keel, false-keel, dead-wood, stem, fore-foot, apron, knight-heads, hawse-pieces, stern-post, siding of the frames, keelson, iron-diagonals, thick-stuff at the floor-heads and futtock-heads, clamps for the beams, shelf-pieces, water-ways, spirketting cat-heads, riding-bolts, hawse-holes, all the wales and outside planking. All the other scantlings are taken to the real breadth of the ship.

If the length in the water-line is from 6½ to 8 times the breadth, divide the length by 6, and consider the quotient as breadth for the above-mentioned scantlings. If the length is more than 8 times the breadth, divide by 7, and if it is more than 9½ times the breadth, divide by 8, and take the scantlings as before.

This is applicable for men-of-war as well as for merchant ships, but for steam-vessels intended only to carry passengers, and for mail-steamers, the scantlings have been made slighter, particularly the moulding of the frames, which according to circumstances has been diminished to something about ¾ of what is put down in the table.

All the dimensions in the table are given in inches, except where otherwise is made known.

	Moulded Breadth in Feet.							
	50	45	40	35	30	25	20	15
Keel, sided	18	16¾	15	13½	12	10½	8½	6¾
„ moulded not less than	18	16½	15	13½	12	11	10	8½
„ below the rabbat not less than	10	9	8	7	6	5½	4½	3¾
„ scarfs, long . . . foot	6	5¾	5	4½	4	3¾	3½	3
Lips of scarfs, thick	3¾	3½	3½	3¼	3	2¾	2½	2¼
Number of bolts in each scarf	8	8	8	7	7	5	4	3
Diameter of bolts in the middle of scarf	1	1	1	1	⅞	¾	¾	⅝
„ „ „ in the lips	¾	⅝	⅝	⅝	⅝	⅝	½	½
False keel, thick	5	4¾	4	3¾	3	2½	2¼	2
Stem sided as keel								
„ outside the outer edge of the rabbat	10	9½	8½	8	7½	7	6½	6
Scarfs of forefoot, long . . . foot	4½	4¼	4	3¾	3½	3¼	3	2¾
Knight-heads sided	18	16¾	15	13½	12	10½	8¾	7½
Plank between stem and knight-head, thick	7	7	6	4¾	4	3½	3	2½
Bolts through knight-heads and stem, diameter	1¾	1¼	1⅛	1	⅞	¾	¾	⅝
Hawse-pieces, number on a side	5	5	5	5	5	3	3	2
„ sided as knight-heads.								
Stern-post sided as keel.								
False-post, thick	5	4¾	4½	4¼	4	3¾	3½	3¼
Wing-transom, up and down	14	12¾	11½	10¼	9	7¾	6½	5¼
„ moulded in the middle	17	15¾	14½	13	11½	10	8	6
Frames, floor timbers, sided, at least	13½	12¼	11	9¾	8½	7¼	6	4¾
1st futtock as floor timbers; the other futtocks diminishing upwards to	11	10¼	9½	8¾	7¾	6¾	5½	4½

		Moulded Breadth in Feet.							
		50	45	40	35	30	25	20	15
Toptimbers sided		10½	9¾	8¾	7¾	6¾	6	4½	4
Frames, moulded at cutting down	from	15	14	13	12	10¾	9¼	7¾	6
	to	17½	16	14½	13	11¼	9¾	8	6¼
" " at the middle between the keel and water-line	from	12½	11	9¾	8½	7½	6½	5½	4½
	to	14½	12	11¼	10¼	8¾	7½	6	5
" " at the load water-line	from	11½	10½	9	7¾	6¾	5½	4½	3½
	to	13	11½	10	8½	7¼	6	4¾	3¾
" " at the main-deck	from	10½	9½	8	7	6	5	4	3
	to	12	10¾	9½	8	6¾	5½	4½	3½
" " at the upper-deck	from	6½	6	5½	5				
	to	7	6½	6	5½				
" " at the sheerstrake frigate .		5½	5	4½	4				
" " " sloop		..	...	5½	5¼	5	4½	3¾	3
" bolts below the water-line square		1⅛	1⅛	1	⅞	¾	⅝	⅝	½
" " above " " "		1	1	⅞	¾	¾	⅝	½	⅜
Kelson, sided as keel.									
" moulded in the middle . .		18	16¾	15	13½	12	10½	8½	6¾
" " at the ends . . .		13½	12½	11¼	10	9	7¾	6¼	5¼
" scarfs, long	feet	3½	3½	3	2¾	2½	2¼	2	1¾
" bolts through the keel,	diam.	1¼	1⅜	1¼	1⅛	1	⅞	¾	⅝
Iron diagonals or riders . .	breadth	6	5½	5	4½	4	3¾	3½	3
" " . . .	thickness	1¼	1⅛	⅞	¾	¾	⅝	½	⅜
Limber-strakes, thick-stuff . .	thick	6½	6	5½	4¾	4½	3½	3	2½
At floorheads and futtock heads	do. at ends	5	4½	4	3½	3	2¾	2¼	2¼
Orlop or lower deck clamps . .	thick	7	6½	6	5	4	3½	3	2½
Footwaling		4	3¾	3½	3¼	3	2½	2¼	2
Hooks and crutches, iron, breadth .		6	5¾	5½	5⅜	5¼	4¾	3¾	3½
" " thickness in middle		4	3½	3¾	3⅜	3¼	3	2½	2
" " " at ends. .		1¼	1⅛	1	1	1	¾	¾	⅝
Number of bolts in each		9	9	9	7	7	7	7	5
Diameter of bolts		1¼	1¼	1¼	1⅛	1	1	⅞	¾
Shelf-pieces, Orlop or lower deck, up, down		11½	11	10½	10	9⅝	8¾	8	
" . . breadth at upper side		13	12½	12	11	9¾	8	4	
" " at the ends .		10	9½	9	8	6¾	6	3	
" " at lower side		8½	8	7½	6½	5½	4½	3	
" bolts through ship's side, diam.		1⅛	1	1	⅞	¾	⅝	⅝	
" *Main-deck*. . up and down		11½	11	10½	10	9½	8¾	8	7½
" " breadth at upper side		13	12½	12	11½	10½	9¾	9	8
" " " at the lower side		10	9½	9	8½	8	7½	7	6
" " breadth at the ends, ⅔ of breadth at the middle.									

	Moulded Breadth in Feet.							
	50	45	40	35	30	25	20	15
Shelf pieces, Upper-deck, up and down .	7	7	7	7				
,, ,, breadth at upper side .	10	9½	9	8½				
,, ,, ,, at lower side .	7	6½	6	5½				
,, ,, bolts through ship's side	⅞	¾	¾	¾				
Main-deck clamps, number of strakes .	3	3	3	3	3	3	2	2
,, thickness of ,,	7½	6½	5¾	5	4½	3½	3	2½
,, ,, at the ends	5½	5	4½	4	3¼	2¾	2½	2
Beams. Lower deck, up and down, fir	12	11	10	9	8	6	5	
,, ,, oak	10½	9½	8½	6½	5	4	3	
,, ,, breadth	13	12	11	9	8	6¾	5½	
,, Main deck, up and down, oak	15	13½	12	10	8½	7	5½	4
,, ,, breadth . .	15½	14	13	11½	10	8½	7	5
,, ,, scarfs, long, at least, ft.	10½	9½	9	8				
,, [Upper deck, up and down	9½	9	8½	8				
,, ,, breadth . . .	10½	10¼	10	9½				
,, ,, scarfs, long, at least, ft.	7	7	7	6½				
Beams, Poop and Forecastle, up and down	...	..	...	4	3½	3		
,, ,, . . . breadth	...	...	..	6½	6	5½		
,, *Rounding up*, Main-deck . .	8¾	8	7½	6½	5½	4¾	4	3¼
,, ,, Upper-deck . .	8½	8	7	6½				
,, ,, Poop and Forecastle	...	...	...	8¾	8¼	8		
Waterways. Lower-deck, breadth .	10	9½	9	8½				
,, to be cut down over beams	2½	2¼	2	2				
,, Main-deck, breadth . .	14½	14	13½	12¾	12	11½	10½	9¼
,, to be cut down over beams	4	3¾	3½	3¼	3	2¾	2¾	2¼
,, Upper deck, breadth .	10½	10½	10½	10½				
,, to be cut down over beams	2¾	2¾	2½	2½				
,, Poop and Forecastle, bdth.	...	...	...	10	10	9½		
,, to be cut down over beams	...	...	..	2	2	1¾		
,, Bolts, lower deck, diam. .	1	1	⅞	⅞				
,, ,, main-deck ,,	1¼	1⅛	1	⅞	¾	⅝	½	⅜
,, ,, upper-deck ,,	¾	¾	¾	¾				
Spirketting. Main-deck . . thick	6	5½	5	4½	4	3½	2¾	2
,, Upper-deck . . ,,	3½	3½	3¼	3¼				
Carlings. Lower deck . square	8	7½	6¾	6	5	4½	3½	
,, Main deck, up and down .	10	9	8	6¾	5½	4¾	3¾	2¾
,, ,, at the ship's side, bdth.	10	9	8	6¾	5½	4½	3¾	2¾
,, ,, midships . ,,	12	10¾	9½	8½	7	5¾	4½	3½
,, ,, at the hatches ,,	16	14¼	12½	10¾	9	7½	6	5
,, ,, Mast-partners ,,	16½	15¼	13¾	12	10¼	8¾	7	5½
,, ,, ,, up and down	14	12½	11¼	9¼	7¾	6½	5	3½
,, Upper-deck, up and down .	6	5¾	5½	5				
,, ,, at the ship's side, bdth.	6	5¾	5½	5				
,, ,, midships ,,	7	6¾	6½	6¼				
,, ,, at the hatches ,,	9½	9	8½	8				

	Moulded Breadth in Feet.							
	50	45	40	35	30	25	20	15
Carlings, Upper-deck, Mast-partners „	10½	10½	10	9½				
„ „ „ up and down	8½	8¼	8	7½				
Ledges. Main-deck, . . breadth	11	10	9	8	7	6	5	4
„ „ . . up and down	5½	5	4¾	4½	4¼	3¾	3	2½
„ Upper deck, . . breadth	6½	6	5½	5				
„ „ . . up and down	4	4	3¾	3½				
Riding bitts, number of pairs . .	2	2	2	1	1	1	1	1
„ thickness . . square	20	18	16	14	12	10	8	6
„ bolts through the beams, diam.	1¼	1⅛	1	⅞	¾	⅝	½	⅜
Topsail sheet-bitts, main and fore, square	12½	11½	10	8¾	7½	6¼	5	
„ mizen . . „	9½	8¾	8					
Bowsprit partners, opening between them	22	19½	17	15	13	10½	8	5½
„ . . . . square	15½	14½	13	11¾	10¼	9¾	8	6¾
„ bolts through main deck beams · di.	1¼	1⅛	1	⅞	¾	⅝	½	⅜
„ bolt through the bowsprit, diam.	3	2¾	2½	2¼	2	1¾	1½	1
Jeer-bitts, main and fore, fore and aft .	11½	10¾	10	8¾	7¾	6½	5½	
„ . . . athwartships	23	21	19½	17¾	15	13	10½	
„ mizen, fore and aft . .	9	8½	8					
„ . . . athwartships	18	17	16					
„ cross-pieces, up and down .	6	5½	5	4½	4	3½	3	2½
„ . . . . breadth	8½	8	7½	7	6½	6	5½	5
„ bolts . . diameter	⅞	¾	⅝	⅝	½	½	½	⅜
Combings and headledges, main-deck, height above deck	12	12	12	10½	10½	9	8½	8
„ „ . thickness	5	5	5	4	4	3½	3½	3
„ „ upper deck, height above deck	4	4	4	4				
„ „ . thickness	4	4	4	4				
Plank of the deck, lower deck, thick .	2½	2½	2½	2½	2¼	2	2	
„ main-deck „	4	4	4	3½	3½	3	3	2½
„ upper deck „	3	3	3	3				
„ poop and forecastle .	...	...	...	2½	2¼	2		
Pillars under main-deck beams athwartships	8	7½	6½	5½	4½	4	4	3½
Pillars under main-deck beams fore and aft	10½	9½	8½	7½	6	4¾	4	3½
Knees, of iron, lower deck . breadth	4½	4	3¾	3½	2¾	2¾	2	
„ . thickness at throat	3¾	3½	3¼	2¾	2⅝	2	1⅝	
„ . „ at the ends	1	⅞	⅞	¾	⅝	½	⅜	
„ length on the beam, feet	3½	3¼	3	2¾	2½	2	1¾	
„ bolts at the throat, diam.	1¼	1⅛	1	⅞	¾	⅝	½	
„ the other bolts . „	1	⅞	¾	⅝	½	⅜		
„ main-deck . breadth	5½	5	4½	4	3½	3	2½	2
„ „ thickness at the throat	4½	4¼	3¾	3½	3	2½	2¼	1½

	Moulded Breadth in Feet.							
	50	45	40	35	30	25	20	15
Knees, of iron, main-deck, at the ends .	1⅛	1	1	⅞	¾	⅝	½	⅜
" length on the beam, feet	4	3¾	3½	3¼	3	2½	2	1¾
" bolts at the throat, diam.	1⅜	1¼	1⅛	1	⅞	¾	⅝	½
" the other bolts . "	1⅛	1	1	⅞	¾	⅝	½	⅜
" upper deck . breadth	3½	3¼	3⅜	3⅛				
" thickness at the throat	3	3	2⅞	2¾				
" " at the ends	¾	¾	⅝	½				
" length on the beam, feet	3	3	2¾	2½				
" bolts at the throat, diam	1	1	⅞	⅞				
" the other bolts . "	¾	¾	¾	¾				
Port-sill, upper, main-deck, up and down	8	8	7½	7				
" lower " "	4	4	3¾	3½	3¼	3	3	2¼
" " upper deck "	3	3	3	3				
Gunwale, frigate . . . thick	3	3	3	3				
" sloop . . . . "	...	...	...	5½	5	4½	3½	2¾
Cat-heads . . . . . square	18	16½	14½	13	11	9	7½	6½
" supporter sided . . .	13½	12½	11	9½	8½	7	6¼	4½
Plank of bottom, oak . . thick	5	4½	4	3½	3	2¾	[illegible]	2¼
" fir . . . . "	5½	5	4½	4	3½	3¼	2¾	2½
Main-wales . . . . . . "	8¼	7½	6¾	6	5¼	4½	3¾	3
" . . . . at the ends	6½	6	5½	5	4	3½	3	2¼
Sheer-strake	5	4½	4½	4¼				
" . . . . at the ends	4	3¾	3½	3¼				
Channels Fore and main } thick at the ship's side	8	7½	6½	5¾	5	4¼	3½	2¾
" . . outer edge	4¼	4	3¾	3½	2¾	2½	2¼	2
bolts through the ship's side, diam.	1¼	1	1	⅞	¾	¾	⅝	⅝
Mizen . . thick at the ship's side	6½	6	5½	4½	4			
" . . . " outer edge	3¾	3½	3½	3¼	2¾			
bolts through the ship's side, diam.	1¼	1	1	⅞	¾			
Hawse-holes, number	4	4	4	4	4	2	2	2
" . . interior diameter	18¾	17	15¼	13½	11½	9¾	8	6
" thickness of iron at the under side	2	1¾	1½	1½	1¼	1	1	¾
" " " upper side	1⅛	1	1	1	1	¾	¾	½
" bolts . . diameter	1¼	1	1	1	1	¾	¾	½
Rudder, breadth about . . feet	5½	4¾	4 4/12	4	3½	3	2¾	2
" head . . . . diameter	25½	23¼	21	18½	16	13½	11	8½
" " . diameter, minimum	23	21	19	16½	14	11½	9	6½
" pintles . . . . diameter	3½	3¼	2¾	2½	2¼	1¾	1⅝	1⅛
" " . . . . length	16	15	14	12¾	11½	9¾	8	6
" " . . . . number	6	6–5	5	5–4	4	4–3	3	3–2
Tiller, of iron, length about . feet	11	10	9	8	7	6	5	3½
" square at rudder-head	5⅛	4⅝	4½	4⅛	3¾	3¼	2⅝	2⅝
" breadth at the fore end	3½	3¼	3	2¾	2½	2¼	1⅝	1½
" . . . . up and down	2¼	2⅛	2¼	2	1⅞	1⅝	1¼	1¼

DIAMETER OF BOLTS FOR WALES AND PLANK OF THE BOTTOM.

Thickness of wales or plank	12–11	10 9–8–7	6	5½–5–4½–4	3½–3	2½–2	1½
Diameter of bolts	1¼	1	⅞	¾	⅝	½	⅜

When two pieces of timber are to be bolted together, the diameter of the bolts may generally be $\frac{1}{14}$ of the thickness of the timber, and this diameter is to be increased by $\frac{1}{16}$ of an inch for every timber the bolt has to go through more than two.

When the length of the tiller is L, the diameter of the cylinder of the steering wheel is 0·152 L; but from this diameter must be taken one diameter of the rope, or a spiral be cut out for the rope. The length of the cylinder of the wheel is eleven times the diameter of the rope. It must be possible to turn the rudder 32 to 35 degrees to each side. Diameter of steering wheel from $\frac{1}{8}$ to $\frac{1}{7}$ of the breadth of the ship.

When the thickness of the keel will allow it, the keel may be tapered to both ends, so that it is sided in the fore-end at the forefoot $\frac{11}{12}$, and in the after-end at the stern-posts $\frac{5}{6}$ of the siding in the middle. The stern-post at the height of the rudder-hole, and the stem at the height of the main-wales are then sided the same as the keel in the middle; but the top of the stem may be $\frac{1}{7}$ more if the timber will allow it.

The stem may be joined to the keel in the same manner as the stern-post, which is as strong as with a forefoot, and this will especially be profitable in large ships, for which a good forefoot may be difficult to procure. On the drawings Plate III. and IV. the stem is supposed to be joined to the keel in this manner.

§ 2. *Diameters of Masts and Spars.*

The main diameter is: for lower masts at the upper deck, for bowsprit at the stem, for topmasts at the lower cap, for topgallantmasts at the topmast cap, for the jibboom at the bowsprit cap, for all yards in the middle, for gaffs at the inner end and for spanker and main-booms at the taffrail.

The main diameter is calculated as follows :—

Main and *Foremasts* made of pieces, one inch in diameter for every 3 to $3\frac{1}{4}$ feet of the whole length.

Mizenmast, $\frac{7}{8}$ of the diameter of the mainmast.

Masts made of one piece, one inch in diameter for every $3\frac{1}{2}$ to $3\frac{3}{4}$ feet of the whole length.

Bowsprit, up and down equal to the mainmast, and athwartships equal to the foremast.

Main and fore-topmast	1 inch for every	3 to $3\frac{1}{4}$	feet of the whole length.
Mizentopmast	1 ,, ,,	$3\frac{1}{4}$ to $3\frac{1}{2}$	
Topgallantmasts	1 ,, ,,	$3\frac{1}{4}$ to $3\frac{1}{2}$	
Royals	1 ,, ,,	$3\frac{3}{4}$ feet of the length.	

above the topgallantmast. When topgallantmast and royal are made in one piece, the main diameter is one inch for every $2\frac{4}{7}$ feet of the length from the fidhole to the stops.

Jibboom, one inch in diameter for every two feet of the length outside the bowsprit cap.

Main and *Fore lower yards*, one inch in diameter for every four feet of the whole length.

Topsail-yards, ⅞ inch in diameter for every four feet of the whole length.

Crossjack-yard, and all *topgallant* and *royal-yards*, one inch in diameter for every 5 feet of the whole length.

Spanker and *main-boom*, one inch in diameter for every 3¼ feet of the whole length.

Gaffs, one inch in diameter for every 3¼ to 4 feet of the whole length.

Studdingsail-booms, one inch in diameter for every 4¼ feet of the whole length.

Studdingsail-yards, one inch in diameter for every 4½ feet of the whole length.

Proportion of the diameter of the head, foot and yard-arms to the main diameter.

	Head.	Foot.
Main and foremast	⅞	⅔
„ „ on schooners .	½	
Mizenmast	⅜	⅓
Bowsprit	4/7	3/7
Topmast	¼	1 at the lower cap.
Topgallantmast and royal in one piece	¼	1 at the topmast cap.
Jibboom	⅜	1 at the bowsprit cap.

	Yard-arms.
Lower yards, with shives	¼
„ „ without shives	⅜
Topsail-yards	⅜
Topgallant and royal-yards	⅜
Spanker or main-boom . . . . ⅔ at the outer end.
„ „ . . . . ⅝ at the inner end.
Gaff . . . . ¼ at the outer end.
Studdingsail-booms . . . . ⅝ „ „
„ yards . . . . . ¾ „ „

The tops are made so broad that the topmast shrouds, if lengthened down, would meet the side of the ship at the channels. Fore and aft the tops are ⅔ of their breadth. By this the length of the cross-pieces and tresseltrees is determined.

Depth of the tresseltrees 1¼ in. for every foot of their length, and the breadth equal to half the depth. Distance between the lower tresseltrees ⅔ of the main diameter of the lower mast.

Breadth of cross-pieces ⅔ of the depth of the tresseltrees, and the depth equal to the half of the breadth. To this depth is added 1 to 1½ inch for the cutting down into the tresseltrees.

On the topmast is:

The thickness of the lower end of the topmast athwartships equal to the distance between the lower tresseltrees.

Fidhole: height ⅔ of the main diameter of the topmast, and width $\frac{3}{16}$ of the same diameter.

Depth of the topmast tresseltrees, one inch for every 3¼ feet of the length of the topgallantmast from the fidhole to the stops. The thickness of topmast tresseltrees $\frac{7}{13}$ of their depth.

Cross-pieces square ⅔ of the depth of the tressel-trees. Breadth of outer ends ¾, and thickness up and down of the ends ½ of the thickness in the middle.

Length of caps, 1¾ times the main diameter of the mast to which it belongs. The breadth of the cap equal to ½ of its length, and the depth equal to ⅓ of its breadth. Breadth of the iron ring round the cap ¾ of the depth of the cap, and thickness of the iron ring ⅓ of its breadth.

3. *Dimensions of the principal Materials for Sea-going Iron Ships.*

	Main Breadth of Ship.							
	50	45	40	35	30	25	20	16
Keel, height . . . inches	12	11½	11	9½	8	7	6	5½
" thickness . . . "	3¾—2¼	3—2	3—2	3—2	2¾—1¾	2½—1½	2—1¼	1¾—1
Frames, angle iron, from "	11/16 × 6½ × 4	11/16 × 6 × 4	5/8 × 5½ × 3½	9/16 × 5 × 3½	½ × 4½ × 3	7/16 × 3¾ × 2¾	3/8 × 3 × 2½	5/16 × 2½ × 2½
" " " to "	5/8 × 6 × 3½	5/8 × 5½ × 3¼	9/16 × 5 × 3	½ × 5 × 3	7/16 × 4 × 2¾	3/8 × 3½ × 2½	5/16 × 3 × 2¼	¼ × 2½ × 2
Garboard strake, from . "	[illegible]	[illegible]	[illegible]	[illegible]	[illegible]	[illegible]	9/16	[illegible]
" " to . "	[illegible]	[illegible]	[illegible]	[illegible]	[illegible]	[illegible]	[illegible]	7/16
Strakes from the garboard to the upper part of the bilge, or to the water-line, and sheerstrakes . from	[illegible]	[illegible]	[illegible]	[illegible]	[illegible]	[illegible]	[illegible]	7/16
to	[illegible]	[illegible]	[illegible]	[illegible]	[illegible]	[illegible]	5/16	6/16
From the water-line to the sheerstrakes . . . from	[illegible]	[illegible]	[illegible]	[illegible]	[illegible]	[illegible]	7/16	[illegible]
to	[illegible]	[illegible]	[illegible]	[illegible]	[illegible]	[illegible]	[illegible]	[illegible]
Beams, stringer-plates upon beam ends, hooks, crutches, floor-plates, and keelsons . . . from	[illegible]	[illegible]	[illegible]	[illegible]	[illegible]	[illegible]	7/16	[illegible]
to	[illegible]	[illegible]	[illegible]	[illegible]	[illegible]	[illegible]	[illegible]	4/16
Beams, height for main-deck .	14½	13	11	9½	8	7	6	4½
" " for upper-deck .	9½	9	8½	8				
" " for lower-deck .	12	10½	9	7½	6	5	4	
Bulkheads, thickness of plates .	[illegible]	[illegible]	[illegible]	[illegible]	[illegible]	5/16	4/16	[illegible]
Rudder, diameter at the head .	6¾	6½	6	5	4½	3¾	3	2¼
" diameter at the heel .	4½	4	3½	3	2¾	2½	2¼	2

DIAMETER OF RIVETS.

Thickness of plates, inch	$\frac{9}{16}$	$\frac{7}{16}$	$\frac{1}{2}$	$\frac{1}{4}$	$\frac{11}{16}$	$\frac{13}{16}$	$\frac{1}{3}$	1
Diameter of rivets ,	$\frac{3}{8}$		$\frac{3}{4}$		$\frac{7}{8}$		1	

The rivets, where the seam is to be water-tight, are not to be nearer to the edges of the plates than a space equal to the diameter of the rivets, and not farther from centre to centre than four times their diameter nor less than three times their diameter.

Rivets through the plating and frames, and through the angle-irons on the beams, and generally where not water-tightness is required between the pieces joined together, may be placed so far from each other as eight times their diameter from centre to centre.

The overlapping of the plates, where double riveting is used, ought not to be less in breadth than five times the diameter of the rivets ; and where single riveting is used, not less than three times the diameter of the rivets.

Distance between the frames 15 to 18 inches, and in river steamboats, 20 to 24 inches. Generally in such steamers, the distance between the frames is greater towards the extremities of the vessel than in the middle ; say, in the engine-room, 15 to 18 inches, and fore and aft, 20 to 24 inches.

APPENDIX.

At page 32 it is said that probably the most advantageous sharpness of the after-body will he acquired by placing the midship section about twice the main hreadtb before the after-end of the construction water-line, and this supposition is founded on experience from several fast-sailing ships and smaller craft. Still, in making the investigations leading to this snpposition it was found that in some vessels, considered very good and fast, the midship section or greatest transverse section was placed farther aft, and in others, also considered very fast, farther forward than twice the main hreadth from the after-end of the construction water-line. This led to the idea, by a sort of theoretical investigation, to find at what distance from the after-end the greatest transverse section ought to he placed to ensure the least resistance or cohesion of the water to the after-hody of a ship.

By experiments made in Sweden, hy the celebrated naval constructor, Admiral Chapman, with models of a tolerably large size, it was found that a minimum of cohesion of the water to the after-body was obtained when both sides of the after-end of the model made an angle with each other of 26° 34′ or an angle with the middle line of 13° 17′.

FIG 1.

If ag, Fig. 1, is the middle line, and the angle $g =$ 13° 17′, gf is $= \dfrac{fi}{\text{tang. } 13° 17′} = \dfrac{fi}{0\cdot236}$.

If ad is the greatest transverse section, there must be a rounding or a curved line ei from this section to where the straight line begins, and the straight line gd must be a tangent to this curved line in the point i; ei may be a parabola with its apex in e, and whose equation is $y^2 = px$, and then, when ik is parallel to ag, we have $ke = \dfrac{kd}{2}$. The angle kid is $=$ the angle g $= 13° 17′$ and therefore $kd = 0\cdot236\, ki$, $\dfrac{kd}{2} = ke =$ $0\cdot118\, ki$; af may be proportional to ae, say $af = q \times$ ae, and, when fi is parallel to ad, af is $= ki$ and fi $= ak$; ak is $= ae - ke = ae - 0\cdot118\, ki = ae -$ $0\cdot118\, q \times ae = ae\,(1 - 0\cdot118\, q) = fi$; $fg = \dfrac{fi}{0\cdot236} =$ $4\cdot2373\, fi = 4\cdot2373\,(1 - 0\cdot118\, q)\, ae = (4\cdot2373 - 4\cdot2373$ $\times 0\cdot118\, q)\, ae = (4\cdot2373 - 0\cdot5\, q)\, ae.$

$af + fg = ag = q \times ae + (4\cdot2373 - 0\cdot5\, q)\, ae = ae\,(0\cdot5\, q + 4\cdot2373).$

$a\,g$ is the distance from the greatest transverse section or the midship section to the after-end of the load waterline, and

$$\text{if } q \text{ is } = 2.25,\ a\,g \text{ is } = 5.3623\,a\,e$$
$$\text{if } q \text{ is } = 2\,0\ ,\ a\,g \text{ is } = 5.2373\,a\,e$$
$$\text{if } q \text{ is } = 1.75,\ a\,g \text{ is } = 5.1123\,a\,e$$
$$\text{if } q \text{ is } = 1\,5\ ,\ a\,g \text{ is } = 4.9873\,a\,e$$
$$\text{if } q \text{ is } = 1.25,\ a\,g \text{ is } = 4.8623\,a\,e$$
$$\text{if } q \text{ is } = 1\ \ ,\ a\,g \text{ is } = 4.7373\,a\,e$$

It seems very reasonable to suppose that, when a ship sails, the water has a great tendency to follow the shortest possible lines along the bottom of the ship, and these lines will be the lines drawn in the body plan on the ship's drawing, generally called diagonals; if these are drawn perpendicular, or nearly so, to most part of the frames. If the above-named experiments on the cohesion of the water to the after-body are correct, these diagonals ought consequently to make an angle with their respective midship lines as nearly as possible equal to 18° 17′, but every one familiar with shipbuilding will instantly see that it is not practicable that every one of them in every place can have this constant inclination to their midship lines; therefore the constructor only can endeavour so to construct the after-body, that as great a part as possible of it may partake of this form. By the parabolical system of construction all the diagonals will necessarily be curved lines, and consequently a small part only of each diagonal will have exactly the inclination towards its midship line that is found by the experiments to give the least cohesion; but as the same experiments show that a small deviation from this inclination not materially affects the cohesion, the after-body may still be

an extremely good one, although perhaps not the very
best one, if only the place of the midship section, in
regard to the after-end of the ship, is so fixed that the
diagonals for medium have the above-named inclination
towards their respective middle lines.

To effect this, let us suppose that the midship section
b e c, Fig. 2, is a parabola
with its apex in the load
water-line *a b,* which in some
ships is the case, and in a
great number of ships very
nearly so. Let the exponent of this parabola be $= u$,
$a b$, Fig. 2, the half-breadth $= \frac{B}{2}$, $a c$ the depth from
the load water-line to the rabbat on the keel $= d$, the
area of the midship section $= \phi = m\, \text{B}\, d$, then $\frac{\phi}{\text{B}}$ is $=$
$m d$, and call $m d = h$. If d is $= m'$ B we have $h =$
$m m'$ B. Draw $b g$ parallel to $a c$, put $b g = h$, draw $g e$
parallel to $a b$, and $e f$ parallel to $a c$; then $f e$ is $= h$
and the diagonal $a e$ drawn from e to the point a in the
water-line, where it intersects the vertical midship line
$a e$, may be considered to be the diagonal, that in most
part of its length from the midship section to the after-
end of the load water-line shall have the inclination of
13° 17' towards the middle line of the same water-line.
In fact $a c$ in the body plan is the same length as $a e$ in
Fig. 1., that represents the plan of this diagonal for the
after-body.

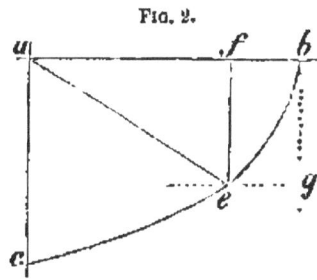

F

As $b\,e\,c$ is a parabola we have $a\,c'' = p \times a\,b$, p being the parameter of the parabola, that is $d'' = \dfrac{p\,\text{B}}{2}$, d is $=$ $m'\,\text{B}$, therefore $d'' = m''\,\text{B}'' = \dfrac{p\,\text{B}}{2}$, and $p = \dfrac{2\,m''\,\text{B}''}{\text{B}}$, $h = m\,m'\,\text{B}$, $h'' = m''\,m''\,\text{B}''$ and $\dfrac{h''}{p} = \dfrac{m''\,m''\,\text{B}''\,\text{B}}{2\,m''\,\text{B}''} = \dfrac{m''\,\text{D}}{2}$.

In the parabola $b\,e\,c$ we have $f\,c'' = p \times f\,b$, that is $h'' = p \times f\,b$ and $\dfrac{h''}{p}$ which is $= \dfrac{m''\,\text{B}}{2} = f\,b$, $a\,f = a\,b - f\,b = \dfrac{\text{B}}{2} - \dfrac{m''\,\text{B}}{2} = \dfrac{\text{B}}{2}(1 - m'')$.

$$a\,f^2 = \frac{\text{B}^2}{4}(1 - m'')^2.$$

$$e\,f^2 = k^2 = m^2\,m'^2\,\text{B}^2 = \frac{4\,m^2\,m'^2\,\text{B}^2}{4}$$

$$a\,f^2 + e\,f^2 = a\,e^2 = \frac{\text{B}^2}{4}\left((1 - m'')^2 + 4\,m^2\,m'^2\right) \text{ and}$$

$$a\,e = \frac{\text{B}}{2}\sqrt{(1 - m'')^2 + 4\,m^2\,m'^2}.$$

The distance from the midship section to the after-end of the load water-line is according to the foregoing, Fig. 1, $= a\,g = a\,e\,(0.5\,q + 4.2373)$ and if the value of $a\,e$ is put in, we have $a\,g = (0.5\,q + 4.2373) \times \dfrac{\text{B}}{2}$ $\sqrt{(1 - m'')^2 + 4\,m^2\,m'^2}$.

The area of the midship section is $= \dfrac{n}{n+1}\,\text{B}\,d$, but it is also $= m\,\text{B}\,d$, therefore $\dfrac{n}{n+1} = m$, and $n = \dfrac{m}{1 - m}$.

If this is to be applied to the frigate, whose elements

of construction are calculated at page 47, we have $m =$ 0·8, $m' = 0·4$, and $n = \dfrac{m}{1-m} = 4$.

The distance from the after-end of the load water-line to the greatest transverse section, or to the after-end of the parallel piece, is then found, when q is = 1·75, to be 2·226 B = 100·6 feet;

When q is 1·5, to be ... 2·171 B = 98·18 feet;
When q is 1·25, „ ... 2·117 B = 95·72 feet;
When q is 1·0, „ ... 2·062 B = 93·26 feet.
On the draught this distance is = 96·73 foot.

In order to see how this theory agrees with practice, the length of the diagonal $a\,e$ in the body plan is found on the drawings for some vessels considered to sail uncommonly well, and then the distance from the after-most perpendicular to the greatest transverse section is found in proportion to $a\,e$. These vessels have been the following, and the above-named distance found as put down for each vessel:—

English Frigate *Phaëton*, constructed by Mr. White, Cowes . 5·1 $a\,e$
Swedish do. *Désirée*, constructed by Capt. Carlsund, naval constructor in the Swedish Navy . . . . . 5·1 $a\,e$
Norwegian do. *Desideria*, constructed by the Author . . 4·6 $a\,e$
English do. *Inconstant*, constructed by Admiral Hays . 5·0 $a\,e$
Norwegian Corvet *Nordstjernen* (The North Star), constructed by the Author . . . . 5·3 $a\,e$
French Brig *L'Espiègle* . . . . 5·1 $a\,e$
American Schooner, the yacht *America* . . . . . 4·9 $a\,e$
Norwegian boat *Svalen* (the Swallow), the breadth only 12 foet 5·0 $a\,e$

Medium . . . 5·01 $a\,e$

This seems to be a very remarkable coincidence between what may be called theory and practice. It

100   APPENDIX.

confirms the accuracy and correctness of the experiments made by Admiral Chapman, and gives a rational cause for the experience gained after the introduction of steam-power: that in long and narrow vessels the greatest breadth and transverse section shall be abaft the middle of the length, nobody doubting that the sharper the fore-body conveniently can be made the less will be its resistance to the water.

THE END.

BRADBURY AND EVANS, PRINTERS, WHITEFRIARS.

THE FOLLOWING ARE

WORKS ON NAVAL ARCHITECTURE.

PUBLISHED BY MR. WEALE.

1.

In 1 volume, 4to. text, and a large atlas folio volume of plates, half bound,
Price 6l. 6s.

THE ELEMENTS AND PRACTICE OF NAVAL

ARCHITECTURE; or, a TREATISE ON SHIP BUILDING, theoretical
and practical, on the best principles established in Great Britain; with
copious Tables of Dimensions, Scantlings, &c. The third edition, with
an Appendix, containing the principles and practice of constructing
the Royal and Mercantile Navies, as invented and introduced by Sir
ROBERT SEPPINGS, Surveyor of the Navy. By JOHN KNOWLES, F.R.S.
Illustrated with a Series of large Draughts and numerous smaller
Engravings.

LIST OF PLATES.

Perspective of the frame of a 100-gun ship.
Construction of on arch, circles, &c.
Cones.
Capstans, crabs.
Conducting bodies and bars.
Floating bodies

Representation of a flying proa.
Experiments on stability.
Scale of solidity of tonnage and displacements.
Machines for driving and drawing bolts.
Longitudinal section and plan of a 74-gun ship.

Plates of Details.—The following are exceedingly large :—

Construction 1. Draught of a ship proposed to carry 80 guns upon two decks, with details.
————— 2. Disposition of the frame for a ship of 80 guns.
————— 3. The planking expanded of the 50-gun ship.
————— 4. Profile of the inboard works of the 80-gun ship.
————— 5. Plans of the gun deck and orlop of ditto.
————— 6. Plans of the quarter deck, forecastle, and upper deck of ditto.
————— 7. Main gear capstan of an 80-gun ship, windlass, &c., and details
————— 8. Midship section of a 74-gun ship; midship section of a 74-gun ship, as proposed by Mr. Snod-

grass; midship section of a 36-gun frigate; midship section of a 36-gun frigate, as proposed by Mr. Snodgrass; sketches of a new plan proposed for framing ships, and of the best mode of adopting iron-work in the construction, and other details.
Construction 9. Sheer draught and plans of a 40-gun frigate, with launch, &c.
—————10. Sheer draught, half-breadth and body plans of a sloop of war.
—————11. Draught of the "Dart" and "Arrow" sloops, as designed by General Bentham.
—————12. A brig of war, 18 guns.
—————13. Inboard works of ditto
—————14. Plans of the upper and lower decks and platforms of a brig of war.

Construction 15. Yacht "Royal Sovereign."

—————16. Yacht built for the Prince Royal of Denmark.

—————17. Plans and Section o. the interior of a fire-ship.

—————18 Draught and plans of a bomb vessel.

—————19. A cutter upon a new construction, with mode of fitting sliding keels.

—————20. Sheer draught, half-breadth and body plans of an East Indiaman.

—————21. Sheer draught, half-breadth and body plans of a West Indiaman.

—————22 A collier brig of 110 tons.

—————23. A Virginia-built boat fitted for a privateer.

—————24. A fast sailing schooner.

—————25 A Virginia pilot boat.

—————26. A Berwick smack.

—————27. A sloop of 60 tons in the London trade, particularly distinguished for her capacity and velocity.

—————28. A Southampton fishing hoy.

—————29. The long boat of an 80-gun ship, showing the nature and construction by whole moulding.

Construction 30. A launch, pinnace, eight-oared cutter, yawl, &c.

—————31. Wherry, life-boat, whale boat, a gig, a swift rowing boat.

—————32. Laying off, plan of the fore body, sheer, and half-breadth plan of the fore-body, belonging to the square bodies, &c.

—————33. Plan of the after-body, sheer, and half-breadth plans of the after-body, &c.

—————34. After-body plan, fore-body plan, sheer and half-breadth plans of the after-cant body, sheer, and half breadth plans of the fore-cant body.

—————35. Horizontal transoms, cant transoms, sheer plan, body plan, &c.

—————36 Square tuck, body plans, sheer and half-breadth plans.

—————37. Hawse pieces, cant hawse pieces, &c.

—————38. Laying off of the stern, laying off of the harpins, plan of the stern, sheer plans, body plans.

—————39. Plans, elevations, and sections of the different contrivances for fitting the store-rooms, &c. on the orlop of an 80-gun ship, showing the method of fitting all ships of the line in future.

',

The plates in large atlas folio, price 2l. 2s.

NAVAL ARCHITECTURE; or, THE RUDIMENTS AND RULES OF SHIP BUILDING; exemplified in a Series of Draughts and Plans, with Observations tending to the Improvement of that important Art. By MARMADUKE STALKARTT, N.A.

THE PLATES CONSIST OF—

1. A long boat for a third-rate, six figures of various draughts
2. A yacht of 142 tons, ten figures of several draughts.
3. A sloop, 331 tons, sheer draught and bottom, fore and aft-bodies.
4 A sloop of war, cant timbers.
5. The bottom and top side.
6. 44-gun frigate, fore and aft, and bottom, a very fine and large engraving.
7. Draughts, several.
8. The shift of the planks in the top

side, and the dispositions of the timbers in ditto.
9 74-gun ship, sheer draught and bottom, fore and aft-bodies.
10. Draughts, several.
11. Right aft, a level view of the stern of a 74-gun ship, side view of the head and quarter-gallery, &c.
12 A cutter, draughts, &c.
13. Exact method of ending the lines of different sections.
14. A frigate, sheer draught, bottom, fore and aft-bodies.

These plates exhibit fineness and correctness of drawing and engraving, and upon a large scale of rare occurrence.

III.

In 3 parts or volumes, in 12mo, with numerous plates and woodcuts, price 3s.

RUDIMENTS OF NAVAL ARCHITECTURE; or, AN EXPOSITION OF THE ELEMENTARY PRINCIPLES OF THE SCIENCE, AND THE PRACTICAL APPLICATION TO NAVAL CONSTRUCTION, for the Use of Beginners. By JAMES PEAKE, Naval Architect.

THE
SERIES OF EDUCATIONAL WORKS

ARE ON SALE IN TWO KINDS OF BINDING,

HAMILTON'S Outlines of the History of England, 4 vols. in 1, bound in cloth . 5s
———— Ditto, in half morocco, gilt, marbled edges . , , . . 5s. 6d
LEVIEN'S History of Greece, 2 vols. In 1, bound in cloth . , , . 3s 0d.
———— Ditto, in half morocco, gilt, marbled edges . . . . . . 4s.
———— History of Rome, 2 vols. in 1, bound in cloth . . . . 3s. 6d.
———— Ditto, in half morocco, gilt, marbled edges . . . . . 4s.
CHRONOLOGY of Civil and Ecclesiastical History, Literature, Art, &c., 2 vols. in 1, bound in cloth . , . . , . . , , , . 3s. 6d.
———— Ditto, in half morocco, gilt, marbled edges. . . . , , . 4s.
CLARKE'S Dictionary of the English Language, bound in cloth . . , 4s. 6d.
———— in half morocco, gilt, marbled edges . . . . , . . . 5s.
———— bound with Dr. Clarke's English Grammar, in cloth . . . 5s 6d.
———— Ditto, in half morocco, gilt, marbled edges . . . . . 6s.
HAMILTON'S Greek and English and English and Greek Dictionary, 4 vols. in 1, bound in cloth 5s
———— Ditto, in half morocco, gilt, marbled edges . . , , . 5s. 6d.
———— Ditto, with the Greek Grammar, bound in cloth . . , . 6s.
———— Ditto, with Ditto, in half morocco, gilt, marbled edges . . . 6s. 6d.
GOODWIN'S Latin and English and English and Latin Dictionary, 2 vols. in 1, bound in cloth . . . . 4s 6d.
———— Ditto, in half morocco, gilt, marbled edges . . . , . . 5s.
———— Ditto, with the Latin Grammar, bound in cloth . . , . 5s. 6d.
———— Ditto, with Ditto, in half morocco, gilt, marbled edges . . , . 6s.
ELWES'S French and English and English and French Dictionary, 2 vols. in 1, in cloth . . . . 3s. 6d.
———— Ditto, in half morocco, gilt, marbled edges . . . . . 4s.
———— Ditto, with the French Grammar, bound in cloth . . . . 4s. 6d.
———— Ditto, with ditto, in half morocco, gilt, marbled edges . . . . 5s.
FRENCH and English Phrase Book, or Vocabulary of all Conversational Words, elaborately set forth for Travelling Use, as a Self Interpreter, bound . . 1s. 6d.
ELWES'S Italian, English, and French,—English, Italian, and French,—French, Italian, and English Dictionary, 3 vols. in 1, bound in cloth , . . 7s. 6d.
———— Ditto, in half morocco, gilt, marbled edges . , . . . 8s. 6d.
———— Ditto, with the Grammar, bound in embossed cloth, marbled edges 8s. 6d.
———— Ditto, with Ditto, in half morocco, gilt, marbled edges . . . 9s.
ELWES'S Spanish and English and English and Spanish Dictionary, 4 vols. in 1, bound in cloth . . . . 5s.
———— Ditto, in half morocco, gilt, marbled edges . . , , , 5s. 6d.
———— Ditto, with the Grammar, bound in cloth . . . . . , 6s.
———— Ditto, with Ditto, in half morocco, gilt, marbled edges , . , 6s. 6d.
HAMILTON'S English, German, and French,—German, French, and English,— French, German, and English Dictionary, 3 vols. in 1, bound in cloth . . 4s.
———— Ditto, in half morocco, gilt, marbled edges . . . . 4s. 6d.
———— Ditto, with the Grammar, bound in embossed cloth, marbled edges . 5s.
———— Ditto, with Ditto, in half morocco, gilt, marbled edges . . . 5s. 6d.
BRESLAU'S Hebrew and English Dictionary, with the Grammar, 2 vols. in 1, bound in cloth . . . . 8s. 6d.
———— Ditto, 2 vols. in 1, in half morocco, 9s. 6d.—With vol. 3, English and Hebrew . . , . , . , ... 12s. 6d.

www.ingramcontent.com/pod-product-compliance
Lightning Source LLC
Chambersburg PA
CBHW021943190326
41519CB00009B/1123